Waste Issues

ISSUES

Volume 161

Series Editor

Lisa Firth

Independence

Educational Publishers
Cambridge

First published by Independence
The Studio, High Green
Great Shelford
Cambridge CB22 5EG
England

© Independence 2008

British Library Cataloguing in Publication Data
Waste Issues – (Issues Series)
I. Firth, Lisa II. Series
363.7'28

ISBN 978 1 86168 454 7

Printed in Great Britain
MWL Print Group Ltd

Cover
The illustration on the front cover is by
Angelo Madrid.

CONTENTS

Useful information for readers

Dear Reader,

Issues: Waste Issues

Waste and recycling have become a political hot potato in recent years, with issues such as rubbish collections and the use of supermarket carrier bags hitting the headlines. Britain's record on waste disposal remains poor, with the UK sending more than 22.6 million tonnes to landfill in 2004-5. Could we recycle more, and would this really help the environment in the long term? This books looks at some of the issues.

The purpose of *Issues*

Waste Issues is the one hundred and sixty-first volume in the **Issues** series. The aim of this series is to offer up-to-date information about important issues in our world. Whether you are a regular reader or new to the series, we do hope you find this book a useful overview of the many and complex issues involved in the topic. This title replaces an older volume in the **Issues** series, Volume 111: **The Waste Problem,** which is now out of print.

Titles in the **Issues** series are resource books designed to be of especial use to those undertaking project work or requiring an overview of facts, opinions and information on a particular subject, particularly as a prelude to undertaking their own research.

The information in this book is not from a single author, publication or organisation; the value of this unique series lies in the fact that it presents information from a wide variety of sources, including:

⇨ Government reports and statistics
⇨ Newspaper articles and features
⇨ Information from think-tanks and policy institutes
⇨ Magazine features and surveys
⇨ Website material
⇨ Literature from lobby groups and charitable organisations.*

Critical evaluation

Because the information reprinted here is from a number of different sources, readers should bear in mind the origin of the text and whether the source is likely to have a particular bias or agenda when presenting information (just as they would if undertaking their own research). It is hoped that, as you read about the many aspects of the issues explored in this book, you will critically evaluate the information presented. It is important that you decide whether you are being presented with facts or opinions. Does the writer give a biased or an unbiased report? If an opinion is being expressed, do you agree with the writer?

Waste Issues offers a useful starting point for those who need convenient access to information about the many issues involved. However, it is only a starting point. Following each article is a URL to the relevant organisation's website, which you may wish to visit for further information.

Kind regards,

Lisa Firth
Editor, **Issues** series

** Please note that Independence Publishers has no political affiliations or opinions on the topics covered in the **Issues** series, and any views quoted in this book are not necessarily those of the publisher or its staff.*

ISSUES TODAY
A RESOURCE FOR KEY STAGE 3

Younger readers can also now benefit from the thorough editorial process which characterises the **Issues** series with the launch of a new range of titles for 11- to 14-year-old students, **Issues Today**. In addition to containing information from a wide range of sources, rewritten with this age group in mind, **Issues Today** titles also feature comprehensive glossaries, an accessible and attractive layout and handy tasks and assignments which can be used in class, for homework or as a revision aid. In addition, these titles are fully photocopiable. For more information, please visit the **Issues Today** section of our website (www.independence. co.uk).

Waste in the UK

Information from the Economic and Social Research Council

This ESRC information provides a statistical overview of waste in the UK. It is designed to introduce the topic rather than be a comprehensive summary.

Historical perspectives

The major component of waste at the beginning of the 20th century consisted of dust from coal fires and cinder. The move away from coal fires in the 1960s saw a drop in dust, coal and cinder waste and currently, paper, board, metals, glass and plastics constitute the main bulk of domestic waste.

What is waste?

Waste is defined as the solid leftovers that are disposed of by, or on behalf of the local authority. Waste is either domestic (household) waste or municipal (business) waste or industrial waste. Due to the increase in waste production in the latter half of the 20th century, systems for managing waste have become more sophisticated. The government classifies waste in two ways: hazardous waste that usually has one or more of four characteristics (ignitability, corrosivity, reactivity, or toxicity); and non-hazardous waste.

How much waste do we produce?

In 2004 the UK produced about 335 million tonnes of waste. Figure 1 on the next page shows the estimated proportion produced by each sector. Industrial and commercial waste from the construction and demolition industry constitute the largest pro-portion of waste.

The total amount of UK household waste in 2005/06 was 28.7 million tonnes, down 3 per cent from 29.6 million tonnes in 2004/05. Graph 1 over the page shows how much municipal waste was recycled over the last 20 years. It shows that in 2004/05 on average each person in the UK produced 517kg of waste, of which 78 per cent was not recycled.

In 2004 the UK produced about 335 million tonnes of waste

A fifth of household waste includes paper and card, of which the cardboard could be used in people's gardens as an excellent compost. Approximately 6-8 per cent of household waste comprises glass bottles and jars. The UK's glass recycling rate is currently around 22 per cent, which is low compared to some countries within Europe such as Switzerland and the Netherlands which have rates as high as 80 per cent.

Each year, approximately 50 million tonnes of waste comes from industrial sources, such as the food, drink and tobacco industries' production processes. Hazardous waste represents less than 2 per cent of all waste but due to its hazardous nature, its treatment is difficult and expensive. The 10-year period from 1994 to 2004 saw the production of hazardous waste rise by 50 per cent. Although this shows a rise in hazardous waste produced, the reporting of such waste has also increased.

Did you know...?

Babies' nappies constitute roughly 2 per cent of household waste. This is the equivalent to nearly 70,000 double-decker buses every year.

Where does our waste go?

The majority of waste goes into landfill sites (74 per cent) while 8 per cent is incinerated and 18 per cent is recycled or composted. It is estimated that 600 million batteries are landfilled every year, representing 20,000-40,000 tonnes, with each battery requiring 50 times more energy to produce it than it generates. Every year over 1 million computers are landfilled with only 20 per cent recycled. On average, every household

in the UK uses, and then throws away, one two-litre plastic bottle every day – with plastic taking hundreds of years to biodegrade. Figure 2 below shows the destination of controlled UK waste, excluding quarrying and mineral waste which accounts for roughly 100 million tonnes. Mining and quarrying waste is not under the EU Waste Framework Directive.

The EU Landfill Directive promotes a decrease in landfilling of waste by requiring member states such as the UK to decrease the quantity of biodegradable material to 35 per cent of 1995 levels by 2016. The aim of the directive is to decrease the adverse affects on the environment of landfills.

Problems with managing waste

Landfilling and incinerating waste can place pressures on the environment, e.g. the leaching of nutrients, heavy metals and other toxic compounds from landfills, or the emission of greenhouse gases from landfills and toxins by incinerators.

Additionally, research has been conducted into associated health risks of populations and the risk of adverse birth outcomes living near a landfill site.

Recycling

There is a current drive by westernised societies to minimise waste production by reducing, reusing and recycling products.

Figure 3 below compares the old European Union 15 countries. Portugal has the worst recycling rate of all European countries at only 4 per cent and the Netherlands recycling the most at 64 per cent. The UK managed 18 per cent of its waste without using landfills or incinerators.

There is a current drive by westernised societies to minimise waste production by reducing, reusing and recycling

---

The Household Waste Recycling Act 2003 sets out a target for 2010 for local authorities to offer all their households a doorstep collection of separate recyclable waste.

Jobs from recycling

As of 2002, 17,700 people are employed in the recycling industry during the collection, sorting and reprocessing of material. Waste Watch, an environmental monitoring group, suggests that if the government were to meet its target of recycling 30 per cent of municipal waste production by 2010, then over 35,000 jobs could be created.

Due to the availability of cheap labour and more markets in developing countries, recyclables such as paper and plastic are being sent to places such as China or Indonesia. According to one report, China imports 200,000 tonnes of plastic and 500,000 tonnes of paper and cardboard each year. The recycling industry in the UK currently offers only £50 a tonne for plastic compared to Chinese companies who are buying it for £120 a tonne, thus reducing the UK's recycling rates and depriving the economy of jobs.
27 July 2007

⇨ The above information is reprinted with kind permission from the Economic and Social Research Council, the UK's leading research funding and training agency addressing economic and social concerns. Visit www.esrcsocietytoday.ac.uk for more information or to view references.
© *ESRC*

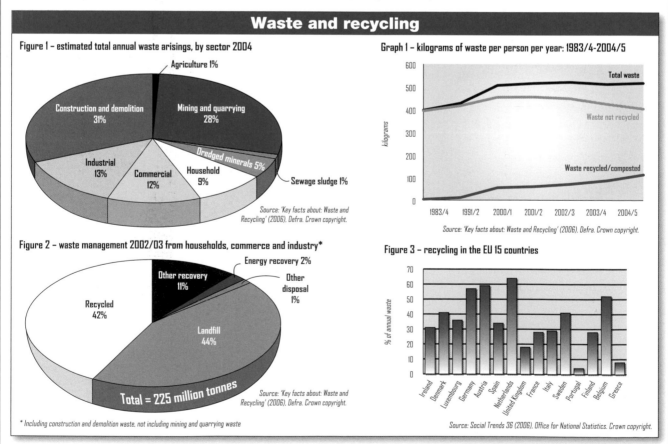

Waste and recycling

Figure 1 – estimated total annual waste arisings, by sector 2004

- Agriculture 1%
- Mining and quarrying 28%
- Construction and demolition 31%
- Industrial 13%
- Commercial 12%
- Household 9%
- Dredged minerals 5%
- Sewage sludge 1%

Source: 'Key facts about: Waste and Recycling' (2006). Defra. Crown copyright.

Graph 1 – kilograms of waste per person per year: 1983/4-2004/5

- Total waste
- Waste not recycled
- Waste recycled/composted

(years: 1983/4, 1991/2, 2000/1, 2001/2, 2002/3, 2003/4, 2004/5)

Source: 'Key facts about: Waste and Recycling' (2006). Defra. Crown copyright.

Figure 2 – waste management 2002/03 from households, commerce and industry*

- Energy recovery 2%
- Other recovery 11%
- Other disposal 1%
- Recycled 42%
- Landfill 44%

Total = 225 million tonnes

Source: 'Key facts about: Waste and Recycling' (2006). Defra. Crown copyright.

* Including construction and demolition waste, not including mining and quarrying waste

Figure 3 – recycling in the EU 15 countries

(% of annual waste; countries: Ireland, Denmark, Luxembourg, Germany, Austria, Spain, Netherlands, United Kingdom, France, Italy, Sweden, Portugal, Finland, Belgium, Greece)

Source: Social Trends 36 (2006). Office for National Statistics. Crown copyright.

Rubbish

Every hour in Britain we throw away enough rubbish to fill the Albert Hall – and most of it ends up in overflowing landfill sites. But how much waste does each of us produce in 24 hours? And what can we do about it?

It was teatime that did it – the third or fourth hot drink of the day, bought in a paper cup with a plastic top, some biscuits in a plastic wrapper, cup and wrapper going not to the bin but to the hoard of detritus growing around my keyboard – and a sudden revulsion. It's not as if I don't recycle. I do, pretty faithfully. But not all of the pile was recyclable, and in some ways that wasn't the point: I was suddenly thinking about just how much of this there must be – per day, per week, per year, per person. And how unnecessary, how thoughtlessly profligate it all was. Obviously the thing to do in such circumstances is rub your nose in your own failings, and so I decided, the following week, to keep all my rubbish for a day; have it photographed, and analysed by experts. Face up to it.

Worries about what we throw away, and how we do it, aren't new. In something like 500BC, Athens moved municipal dumping well away from its city walls; Britain's first dustmen were Romans; we all know, from school lessons or from films, about the cess-filled streets of medieval, even 19th-century London. In 1874, Britain discovered the waste incinerator, and a year later, London acquired a sewerage system; a quarter of a century after that came the widespread use of landfill: all along, as Richard Girling puts it in his book *Rubbish! Dirt on Our Hands and Crisis Ahead*, 'waste policy, as far as there was one, was driven by the politics of disgust'. Now, increasingly, it is driven by fear of the revenge our planet will inflict for our thousands of years of negligence. Driven, too, by sheer scale.

For there are more of us than ever before, and we live more closely packed together. Multiply my plastic cup or two, a sweet wrapper or three, by more than 60 million Britons, every day of every year, add industrial waste,

By Aida Edemariam

electronic waste, hospital waste ... In 2004, according to the Department for Environment, Food and Rural Affairs, the United Kingdom produced about 335m tonnes of waste; every hour we throw away enough to fill the Royal Albert Hall. Last year, 5.5m tonnes of household waste were collected in England alone. Each of Britain's 29m households, says Mark Barthel, an expert in waste reduction for the government-funded Waste and Resources Action Programme (Wrap), throws away the weight-equivalent of one teenage elephant each year. Imagine, he says, how much space 29 million teenage elephants would take up (that is, if they weren't buried in landfill, recycled, or shipped to China). Each of us produces an average of seven times our own weight in waste each year. According to figures for 2004-5 (the latest available from the Environment Agency), the east of England, which traditionally has accepted large amounts of London's waste, had only three years' landfill capacity remaining at current rates of disposal; in London it was four years, and Wales five. The full sites will, and must, be replaced by new ones, but these figures are a graphic illustration of the scale of the problem.

Some of the culprits are well known, and are beginning, slowly, to be dealt with, though in the face of mountains of Styrofoam-cradled electrical goods (not to mention pears), we may feel that a few retailers' gestures on packaging are long, long overdue. But, says Barthel, some packaging is necessary, and retailers have 'this fine balancing act between providing packaging which not only protects the food but also presents consumers with key pieces of information, such as whether it contains nuts, cooking

instructions, storage instructions. I suspect we'll see a lot more produce packed loosely, because that seems to me to be a fairly easy target, though the challenge then for retailers is that the transit packaging is robust enough.' If there is the will, adds Chris Davey from Recycle Now, a national consumer campaign run by Wrap, 'we can design out [of products] a lot of materials that are currently difficult to recycle'.

Every hour in Britain we throw away enough rubbish to fill the Albert Hall

But there are more subtle, rather more difficult things to deal with, too – our attitudes and habits. There is the besetting sin indicated by my sorry little pile, for instance – reaching for convenience. It began with my first meal of the day: a croissant in a paper bag, coffee in a paper cup with a plastic lid, a clear plastic glass of cranberry juice. And, doing what I do, a pile of newspapers. Lunch was soup from the canteen in a cardboard bowl, eaten with a plastic spoon, bread spread (using a plastic knife, of course) with individually wrapped butter. And then there were the Starburst wrappers, biscuit wrappers, sugar sachets, a tea bag ... It was actually better, in some ways, than usual. There is often a polystyrene box from the canteen. And all of it, 24 hours of observation revealed, was acquired at work – at home there are mugs and glasses and pots and plates, and they, of course, can be washed and reused, a point gently but firmly made when my heap was scrutinised by a panel of rubbish analysts: Barthel, Davey and Joy Blizzard, from the Local Authority Recycling Advisory Committee.

'Invest in a mug and a spoon,' said Blizzard. And if you must use a plastic cup and plastic lids, Davey said, why not also use an organisation such as Save a Cup, which will collect them from your place of work and turn them into such things as rulers and pens?

If tea bags end up in landfill, they do biodegrade eventually. But, in the absence of oxygen, this process produces methane, which is 23 times as powerful a greenhouse gas as carbon dioxide (over a period of 100 years); and '38% of the UK's methane emissions come from landfill sites,' says Barthel - obviously a climate change no-no. Instead, he suggests I compost them, which is easier said than done in an office. And I had not realised one of the effects of recycling was to cut CO2 emissions: according to the International Aluminium Institute, recycling aluminium - from drinks cans or foil - is 95% more energy-efficient than making aluminium from raw bauxite. With steel, the metals giant Corus claims, the figure is 75%. According to Wrap, recycling in Britain is now the equivalent of taking 3.5m cars off the road in terms of reducing carbon dioxide emissions.

There are, as I suspected, things in my pile that cannot be dealt with: Starburst wrappers are laminates,

made of different polymers, and because these polymers vary from manufacturer to manufacturer, and are present in such small volumes, there is no market for them as recyclates, quite apart from the challenge of sorting them. This applies to most chocolate bar wrappers and to all crisp packets. Do not be fooled by the fact that many look as if they are made of foil, which is recyclable: all you've go to do is squash them in your hands. If they slowly regain their shape, it's nothing doing.

We cannot pretend that we are not in trouble, and a large part of the solution lies in taking personal responsibility

Some of the trouble begins before we even get to this point. I grew up in Ethiopia, where used plastic bags, cans and bottles were valuable enough to be bought and sold, providing scratched livelihoods in themselves: throwing away a jam jar that might conceivably be used for something still causes a little stab of guilt, guilt that's quieted, but not entirely stilled, by being able to put it in a green box; for those who grew up in plenty, as those born in postwar Britain have done, that guilt must be learned.

Some degree of waste is unavoidable, and no one should pretend otherwise; the torrents of it that we produce,

however, say something about the way we live. 'Junk,' says Girling, 'is not so much a by-product of modern life as the foundation of it. We have a daily diet of junk mail, junk food... Good, workable mobile phones become junk as soon as manufacturers launch their new ranges. Clothes become junk every spring and autumn, as fashion changes its mantle.' (More like every six weeks or so, I would counter.)

When it comes to food, it appears that we are dealing with a kind of epidemic of ineptitude. People do not check cupboards and write lists before they shop, says Barthel (who is working on a campaign to reduce food waste), so they duplicate things and are easily enticed by in-store advertising, buying one to get one 'free'. Food is wasted because it is not stored properly - most fridges are set too high, the average being 6.64C, when it should be between 0C and 5C. About 60% of homes cook too much for each meal, and leftovers get thrown out. Students and young professionals aged 16-35 are more likely to waste food than anyone else. Young families are next in line; a category called, somewhat creepily, social renters - meaning people in housing association or council property - are next, partly because they do not necessarily have freezers or much storage space. According to the Food Standards Agency, only 35% of us understand that a use-by date applies to chilled and perishable food; even fewer, 24%, understand that 'best before' means there may be a reduction in prime quality, not that the product will be dangerous - so we throw out lots of perfectly edible food. Even the most basic home economics skills are atrophying. Increasingly we suffer, in a brilliant phrase used by one of Barthel's research colleagues, from 'food blindness': an inability to look at a cupboard full of ingredients and work out how they might be combined.

When that is added to another sort of blindness - a failure of imagination - the result is obvious. We no longer see waste flowing through the streets and, except in very rural places, we don't have to cart it away ourselves. It goes into a kitchen bin, then to dustbins, and thence out of mind

– until the landfills fill, not in some unimaginable future, but in the next half-decade.

Increasingly, of course, Britain is recycling: 59% of people in England are now what Wrap calls committed recyclers, compared to 45% less than two years ago; but in 2004/05, that accounted for only 26.7% of waste: significant, but still a curiously unimpressive figure.

We cannot pretend that we are not in trouble, and a large part of the solution lies in taking personal responsibility: what each of us does with the waste we produce, and how aware we are that we are doing it. 'The ideal state,' says Davey, 'is where we are only sending for disposal that which cannot possibly be recycled or reused.' I know what I have to do, for starters. Bring a mug to work. And a spoon. And maybe, if I can get slightly more organised, scan my cupboards and make my own lunch.

One day of Aida's rubbish

⇨ Paper coffee cup with plastic top
⇨ Cardboard bowl for soup (there is often a polystyrene box for food from the canteen that I eat at my desk)
⇨ Starburst wrappers
⇨ Newspapers – 10
⇨ Paper bag for croissant
⇨ Cardboard carrier for cups
⇨ Clear plastic cup for fruit juice
⇨ Biscuit wrapper
⇨ 5 colour printouts of photos
⇨ Butter wrapper
⇨ Tea bag
⇨ Paper napkin
⇨ Plastic spoon
⇨ Stirrer
⇨ Sugar wrappers
⇨ Press release for a glow-in-the-dark bra

This article was amended on Sunday 24 June 2007. Methane is 23 times as powerful a greenhouse gas as CO2 (over a period of 100 years) and not 23%, as was stated in the article above. This has been corrected.
11 June 2007
© *Guardian Newspapers Limited 2008*

Wacky waste facts

Information from Waste Online

Have you ever thought about how much rubbish you and your family throw away every week? Or why we need to stop throwing so much of it away? This page is full of amazing waste facts. Did you know that...

General garbage
⇨ The UK produces more than 434 million tonnes of waste every year. This rate of rubbish generation would fill the Albert Hall in London in less than 2 hours.
⇨ Every year UK households throw away the equivalent of 3½ million double-decker buses (almost 30 million tonnes), a queue of which would stretch from London to Sydney (Australia) and back.
⇨ On average, each person in the UK throws away seven times their body weight (about 500kg) in rubbish every year.

Glass
⇨ On average, every family in the UK consumes around 330 glass bottles and jars a year. (British Glass)
⇨ It is not known how long glass takes to break down but it is so long that glass made in the Middle East over 3000 years ago can still be found today.
⇨ Recycling two bottles saves enough energy to boil water for five cups of tea.

Fantastic plastic
⇨ Every year, an estimated 17½ billion plastic bags are given away by supermarkets. This is equivalent to over 290 bags for every person in the UK. 17½ billion seconds ago it was the year 1449.
⇨ We produce and use 20 times more plastic today than we did 50 years ago!

Oil
⇨ 1 litre of oil can pollute 1 million litres of fresh drinking water. (Scottish Oil Care Campaign)
⇨ Waste oil from nearly 3 million car oil changes in Britain is not collected. If collected properly, this could meet the annual energy needs of 1.5 million people. (Scottish Oil Care Campaign)

Preposterous paper
⇨ About one-fifth of the contents of household dustbins consists of paper and card, of which half is newspapers and magazines. This is equivalent to over 4kg of waste paper per household in the UK each week.

Persistent packaging
⇨ In 2001 UK households produced the equivalent weight of 245 jumbo jets per week in packaging waste.
⇨ Every year each person produces 4 times as much packaging waste as their luggage allowance on a jumbo jet.

Revolting rubbish
⇨ Babies' nappies makes up about 2% of the average household rubbish. This is equivalent to the weight of nearly 70,000 double-decker buses every year. If lined up front to end, the buses would stretch from London to Edinburgh.
9 June 2008

⇨ The above information is reprinted with kind permission from Waste Online. Visit www.wasteonline.org.uk for more information.
© *Waste Online*

Waste management

Information from rubbish.co.uk

We have become casual about rubbish in our throwaway world, and leave it to municipal authorities to deal with it. But the pressure on landfill sites demands new solutions, and this means that, increasingly, environmental responsibility for waste disposal will be redirected back towards consumers and manufacturers.

The quantity of rubbish we produce has grown exponentially over the past century, with the relentless rise of consumer goods, packaging and plastics. In the not-so-distant past, when goods were sold in paper bags, and virtually everything was used and re-used, or burnt in the kitchen grate, households would have little to chuck out besides ash. Now, Britain produces 100 million tonnes of rubbish a year. Domestic or household waste accounts for one-sixth of this total – or nearly 300kg for every man, woman and child.

Most of this rubbish ends up in landfill sites, but space is fast running out. With growing environmental sensitivities, and ever more stringent EU directives, and the increasing economic burdens of waste management, the government is having to radically rethink its waste-management strategy.

Waste collection

In Britain, as in all Western industrialised countries, waste collection is the responsibility of the local authorities, and ultimately the government. Efficient and effective waste management is an essential part of civilised society. It is a function that is largely taken for granted: its importance and sheer scale often only comes to notice when things go wrong – for instance during a strike.

Most people have a fair idea of what they can and cannot put in their domestic rubbish bins; it is the duty of the local authority to keep reminding them of this. For larger quantities of rubbish and bulkier items, such as white goods (notably fridges and freezers), the public also needs access to waste disposal sites (or civic amenity sites); but some local authorities will also provide an on-demand service to collect bulky items directly from householders. These authorities are constantly performing a balancing act: failure to give householders, and indeed businesses, fair and free (or low-cost) access to waste disposal will result in fly-tipping and other such abuses, and it can prove even more costly to repair the damage than to provide the necessary services.

All local authorities in Britain publish details of their policies and facilities for waste disposal, and these are accessible through their websites. For a more general view of government policy, see the site of the Department for the Environment, Food and Rural Affairs: www.defra.gov.uk

Hazardous waste

Some kinds of waste must be kept apart from the usual stream of waste disposal. Hazardous waste is any kind of waste that could pose a serious danger to human beings or to other living organisms. It includes such things as chemicals, clinical (or medical) waste, fuel, paint, asbestos, gas canisters, pesticides, car batteries – anything that could be toxic, radioactive, flammable, corrosive or infectious. By far the biggest quantity of hazardous waste is produced by industry – 5 million tonnes of the stuff every year. But individual homeowners also produce their share: if you are in any doubt about how to dispose of such items, consult your local authority: there may well be special facilities at your local civic amenity site. For more details about legislation, directives and regulations concerning hazardous waste, see www.environment-agency.gov.uk

Landfill

What happens to all this waste after it leaves your home? The most common destination for rubbish worldwide is a landfill site. Essentially, the rubbish is tipped into a big hole in the ground – usually a hole created by quarrying or mining. In Britain about 70 per cent of waste goes to landfill sites. Modern landfills are carefully constructed and managed. They have impermeable floors of clay or heavy-duty plastic, and pipes to collect the toxic liquid called leachate, which drains to the bottom, and which

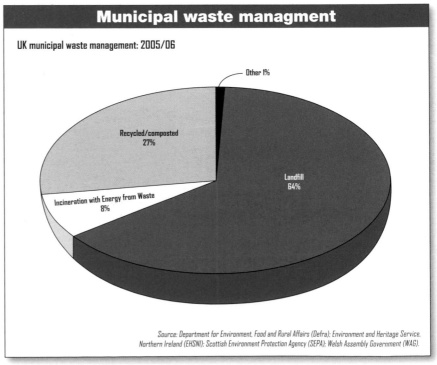

Municipal waste managment

UK municipal waste management: 2005/06

- Other 1%
- Recycled/composted 27%
- Incineration with Energy from Waste 8%
- Landfill 64%

Source: Department for Environment, Food and Rural Affairs (Defra); Environment and Heritage Service, Northern Ireland (EHSNI); Scottish Environment Protection Agency (SEPA); Welsh Assembly Government (WAG).

otherwise might pollute groundwater. Heavy machinery is used to compact the rubbish, and layers of soil may be spread over it to prevent odour, vermin and windblown debris. The decomposition that takes place in a landfill produces flammable, toxic gases, notably methane, an environmentally damaging greenhouse gas. However, this can be flared off through pipes, or alternatively it can be piped to a plant and burnt as a fuel to generate electricity. In some places (notably China) it is piped to households and used as a domestic gas. Once full, landfill sites can be covered over and sealed, and eventually rehabilitated as airports, golf courses, dry ski-slopes, and so on. There are separate landfill sites for hazardous waste, which, needless to say, are treated differently.

Landfill is a comparatively cheap option. The trouble with landfill is that virtually all industrialised countries are rapidly running out of suitable sites. Existing sites are filling up fast, and proposals for new sites are met with insurmountable opposition by those living in the vicinity. Land near big cities is very expensive: cheaper land further away incurs higher transport costs and journey times (collection and transportation already accounts for 75 per cent of waste management costs). From an environmental point of view, landfills are a poor option. The EU wants the UK's use of landfills to be cut to half its 1995 levels by 2013, and to 35 per cent by 2020 – a tough agenda. But by 2020, according to government targets, only 25 per cent of all rubbish will go to landfill. The key to success in this strategy lies in dramatically reducing the volume of waste produced in the first place, largely through recycling (nearly half will be recycled, according to the plan). Meanwhile, the pace accelerates as the total volume of waste in Britain grows by 3 per cent year on year, pushing waste disposal ever closer to crisis.

Incineration

So, if you cannot dump rubbish on the land, why not burn it? Incinerators reduce 85 per cent of rubbish to fine ash. This is what already happens to about 10 per cent of Britain's rubbish. What's more, we can use the rubbish as a fuel to produce electricity. The target is to increase the proportion of rubbish that is incinerated to 27 per cent by 2020.

In Britain about 70 per cent of waste goes to landfill sites

Unfortunately, rubbish incinerators produce a large volume of toxic waste, in the form of gas and solids. Create air pollution, and the environmental fallout becomes an international issue. In recent years, great technical advances have been made in the process of cleaning ('scrubbing') gas emissions from incinerators, but these add considerably to the costs, and still some residual toxic pollutants escape. Another criticism is that incineration destroys large quantities of potentially recyclable materials, so the net energy loss is substantial. And it is difficult to build new incinerators, because – as with landfills – no one wants to live near to one.

The 'three Rs' of rubbish

The search for a solution to waste disposal has produced so-called 'waste hierarchy', or the 'three Rs' of rubbish: Reduce, Re-use, Recycle. We could all be less profligate with our rubbish. In the free-market economy, however, there has to be either an incentive or a penalty – a carrot or a stick. For manufacturers and businesses, the carrot could come from earning 'green' credentials that would attract customers towards them: for instance, some supermarkets have opted to use biodegradable plastic bags, or introduced a no-plastic-bag policy (customers have to remember to bring their own). The stick could be a regime of government enforcement through legislation, penalties and tax.

Zero-waste

Through the policy of Reduce, Re-use, Recycle, bit by bit we edge towards the ultimate goal of producing no rubbish at all: a world of zero-waste. Everything could, in principle, be re-used or recycled – even currently irreducible residues. One initiative in this direction is to make manufacturers responsible for the recycling of their products (household appliances, cell phones, audiovisual equipment and so on) when consumers have finished them; this would quickly force manufacturers to incorporate features into the design of their products to render them easy to dismantle and make recycling cost-effective.

Already EU legislation is pushing manufacturers along this path, such as the Waste Electrical and Electronic Equipment (WEEE) directive, which makes them responsible for financing the collection, treatment, recovery and environmentally sound disposal of electrical equipment. Such developments will significantly change our attitude to rubbish: currently we churn it out and expect the environment to bear the cost; in the future, this cost will be increasingly transferred to manufacturers and the consumers.

⇨ The above information is reprinted with kind permission from rubbish. co.uk. Visit www.rubbish.co.uk for more information.

© rubbish.co.uk

Consumer adultery – the new British vice

In the UK we throw away more consumer products, and faster, than anywhere else in Europe. The result is a shocking – and unsustainable – mountain of discarded hardware

By Lois Rogers

Just as Britain tops the European league for marriage breakdowns, so it also now tops the league for falling in love with consumer products and then throwing them away when newer models come out.

This social trend, which product designers have termed 'adulterous consumption', has given us the biggest methane-producing rubbish tip in Europe – and the biggest headache in deciding what to do with our waste.

Last month, Britain narrowly escaped a huge and embarrassing fine from the European Union by finally implementing the Brussels-inspired Waste Electrical and Electronic Equipment (WEEE) directive. The directive seeks to prevent us from tossing any more of these products with highly toxic, non-degradable components into holes in the ground. From now on, manufacturers of everything from electric toothbrushes and hairdryers to kettles, lighting equipment and washing machines will be forced to take back their products for recycling.

While other European states – notably Germany and the Scandinavian countries – have seen the writing on the wall and have been recycling for more than two decades, Britain's binge borrowers have remained obstinately wedded to their credit cards and have gone on consuming and discarding with abandon.

We were the very last country in Europe to adopt the new law, and the problems with compliance seem insurmountable unless we return to a more conventional loyalty to consumer products.

British women discard their hairdryers after three years, usually in favour of another one that simply looks different. The average ownership of a mobile phone lasts 18 months. And manufacturers of DIY tools have calculated that most power drills are lost or abandoned after a single weekend.

'We don't throw things away because they are broken – it's usually because we have fallen out of love with them,' says Jonathan Chapman, a senior lecturer in design at the University of Brighton, who is trying to promote what he calls 'emotionally durable' design as a way of reducing the generation of toxic waste.

We are producing one million tonnes of electrical wreckage annually, a volume that is rising by 5 per cent year on year

Researchers in the United States have calculated that only 1 per cent of all the materials flowing through their domestic economy goes into products which are still being used six months later. Chapman believes that, without a big shift in our attitude to the things we live with, the UK will soon catch up. 'At the beginning of a relationship with a product, we consume it rampantly,' he says. 'Then consumption becomes routine, and then we stop thinking about it altogether and start noticing newer models. Often the relationship ends because the product is not doing something we want it to do, or it has started doing something we didn't think it would do, but not because it doesn't work. Unless we return to more sustainable relationships with these possessions we are going to have a really huge problem.'

How to square this reckless attitude with the demands of the WEEE directive is a crisis of such proportions that no one has dared look at it.

All manufacturers, importers, distributors and retailers of electrical equipment have until 15 March to register for WEEE compliance. They are required to provide precise information on the weight of the products on the market. They will then have to demonstrate that they are taking back and recycling an agreed quantity.

In practice, they will pay local authorities to provide 'designated collection facilities' (DCFs), which will perform the laborious function of recovering, sorting and returning their products.

The British Retail Consortium has agreed to provide £10m to upgrade local-authority tips to DCFs, but this works out at roughly £6,000 per site – hardly enough to cover the purchase of the separate containers required, let alone wages for the vast armies of additional staff who will be needed to sort copper wire from cadmium and computer screens from lead components.

The only way the directive is likely to work is by somehow engendering a sense of collective social responsibility for waste management, yet there has been no public education campaign, and most consumers are unaware of the existence of the WEEE requirement.

Last month, a county council waste manager in West Sussex admitted that it is not clear where the facilities are to deal with the anticipated WEEE mountains. 'We're not telling the public about this because we don't want them asking questions when we don't know the answers,' she said.

This, coupled with our fragmented waste-disposal industry, and our usual hostility to any anonymous edict handed down from Brussels, is a recipe for chaos.

Others are exasperated at the lack of preparation. 'Britain has known this electrical equipment directive was coming for a good ten to 15 years, but instead of getting properly prepared for it, the general response has been for people to stick their heads in the sand,' says Cerys Ponting, whose work at Cardiff business school on the effects of the WEEE rules is to be published shortly.

Although she has found that most British consumers are oblivious to the implications of our throwaway culture, she does think attitudes will change. 'It is like the early days of seat-belt laws, when people still didn't really see the point of them,' she says. 'Until now there has been little pressure from the government to recycle.

'People somehow regard electronics as a clean industry and they don't understand how much of a problem it actually is. That is slowly changing and there is more and more public understanding of the need to be responsible. We are going to run out of landfill sites and many raw materials fairly soon. Local authorities recognise that, but they are still at the stage of experimenting with different methods of tackling the recycling issue.'

Meanwhile the situation is becoming critical. We are producing one million tonnes of electrical wreckage annually, a volume that is rising by 5 per cent year on year – much faster than the generation of other types of waste.

The average British household contains 25 electrical products, of which at least five are thrown away every year. Two million personal computers are discarded annually. This voracious consumption is being fuelled by plummeting prices. According to the Office for National Statistics, the price of a personal computer has fallen by 93 per cent, in real terms, in the past decade. Prices of televisions, DVD players and vacuum cleaners have fallen by 45 per cent over the same timescale.

Because of an old, and somewhat irrational, resistance in this country to incinerating rubbish, the vast majority of our electronic junk is simply tipped into the disused quarries that conveniently pepper most of Britain's shire counties. Items that may have been used for just a few hours during their working lives are being left to sit underground for thousands of years, giving off copious amounts of methane, a highly potent greenhouse gas that is 23 times more damaging than carbon dioxide.

The proliferating volume of 'large WEEE', as it is known in the industry, presents even more of a problem. Most people find it physically impossible to conceal washing machines and cookers in their dustbins, and there has been a slow move towards recycling the raw materials from such items – or at least not dumping them in holes in the ground.

Government waste advisers fear, however, that the new directive may simply lead to increasing quantities of such discarded goods being exported illegally to countries such as India or Nigeria, where desperate workforces will do just about anything for money, including stripping out heavy metals and other toxic materials from appliances by hand. A 2005 report from the European Commission described an enforcement operation, carried out in 17 European seaports, during which 140 waste shipments were found. Although almost half of these cargoes turned out to be totally illegal, there is no evidence of any major prosecutions. A 15-year-old UN convention, designed to prevent the export of hazardous waste from developed to developing nations, has been similarly ineffective.

Other academics have contrasted the growth of adulterous consumption with our paradoxical attachment to ancient jeans, old teddy bears and worn-out wooden spoons.

Tim Cooper, head of the Centre for Sustainable Consumption at Sheffield Hallam University, says harnessing this desire for connection to our possessions is the key to preventing disaster. He says that the WEEE directive and other legislation restricting the use of hazardous substances in manufactured items should eventually lead to a generation of more durable and repairable items that are not encased in the sealed units which prevent a long life anyway.

'People do like the idea of developing long-term relationships with their possessions; it is just that they have been prevented from doing so by industry, which is geared around stimulating a continuous sense of need for change in order to sell more and more,' he says.

'It is true, though, that there has been no evidence so far of any trend to make things which last longer or which are even more recyclable, and it does look as if things will get considerably worse before they get better.'

Five ways to dispose ethically:
http://www.sofaproject.org.uk
The Bristol-based recycling charity Sofa sells on donated furniture and electrical appliances. Anyone can buy from the charity, but if you are on a low income you get a 25 per cent discount.
http://www.createuk.com
Collects and recycles fridges, freezers, cookers and washing machines. Items suitable for reuse are separated and passed on to refurbishment operations.
http://www.seek-it.co.uk
Computer and software disposal. Good for the security-conscious: Seek-it wipes hard drives. The items it collects are resold or reused as donations to projects in the UK and Africa through www.it-exchange.org
http://www.wasteonline.org.uk
A website that provides information on recommended companies throughout the UK that recycle and reuse electrical goods – from computers to lighting – as well as others specialising in industrial plastics and food.
http://www.actionaidrecycling.org.uk
Collects ink and toner cartridges, mobile phones and PDAs. All are recycled in order to help fund ActionAid's charitable projects in the third world.
5 February 2007

⇨ The above information is reprinted with kind permission from the *New Statesman*. Visit www.newstatesman.com for more information.

Wasteful Britain: the 'dustbin of Europe'

Britain is the 'dustbin of Europe' and could run out of landfill space within nine years as UK households dump more rubbish than any other EU country

By Caroline Gammell

The UK sends the same amount of waste to landfill as the 18 EU countries with the lowest landfill rates combined.

Britain has dedicated an area the size of Warwick – 109 square miles – exclusively to landfill and will run out of space for dumping in less than a decade if current trends continue.

The stark warning came from the Local Government Association, who said the UK sent more than 22.6 million tonnes to landfill in 2004-5, which are the most recent dates comparable with the rest of Europe.

This contrasted with 17.6 million tonnes from Italy, Spain which dumps 14.2 million tonnes, 12 million tonnes from France and Poland which dumps 8.6 million tonnes.

Paul Bettison, from the LGA Environment Board, said: 'Britain is the dustbin of Europe with more rubbish being thrown into landfill than any other country on the continent.

'For decades people have been used to being able to throw their rubbish away without worrying about the consequences. Those days are now over.

'There needs to be an urgent and radical overhaul of the way in which rubbish is thrown away. Local people, businesses and councils all have a vital role to play to protect our countryside before it becomes buried in a mountain of rubbish.'

Landfill levels in the UK dropped significantly over the last 12 months, but Mr Bettison said other countries in Europe had also reduced the amount they discarded, which left the UK on 'the top of the rubbish heap'.

Figures released by the Department for Environment, Food and Rural Affairs (Defra) showed levels of waste sent to landfill in England alone dropped from 17.9 million in 2005-6 to 16.9 million tonnes in 2006-7.

England recycled or composted more than 30 per cent of household waste in 2006-7, compared to 27 per cent in 2005-6.

Mr Bettison welcomed the figures but warned that more needed to be done to avoid being penalised by the EU Landfill Directive.

'It is encouraging to see people doing their bit by recycling more and reducing this country's reliance on landfill,' he said.

'It is pleasing to see our recycling rates exceed 30 per cent for the first time, but the fact remains other countries on the continent are still recycling up to twice as much.

'Councils and council tax payers are still facing fines of up to £3 billion if we do not dramatically reduce the amount of waste thrown into landfill, and so it is vital we look at alternatives to the status quo.'

The directive sets the UK a number of targets, including that by 2020 the amount of biodegradable municipal waste sent to landfill should be no more than 35 per cent of the amount produced in 1995.

12 November 2007

© Telegraph Group Limited, London 2008

Wasted food now costs UK homes £10 billion

Information from WRAP

The cost of needlessly wasted food to UK households is £10 billion a year, £2 billion higher than previously estimated according to new research published today by WRAP.

The research gives detailed new insights into the nature and amount of food waste thrown away in the UK and is believed to be the most

comprehensive study of its kind ever carried out.

It reveals that the average household throws out £420 of good food a year.

For the average family with children it's higher at £610 – money which could have helped pay household bills.

Researchers found that more than half the good food thrown out, worth £6 billion a year, is bought and simply left unused or untouched. For example, each day 1.3 million unopened yoghurt pots, 5,500 whole chickens and 440,000 ready meals are thrown away in the UK. The study revealed that £1 billion worth of wasted food is still 'in date'. It costs local authorities £1 billion a year to dispose of food waste.

Stopping the waste of good food could avoid 18 million tonnes of carbon dioxide equivalents from being emitted each year – the same as taking 1 in 5 cars off UK roads.

In the UK we are throwing away one-third of the food we buy

Launching the report *The Food We Waste* today Liz Goodwin, Chief Executive of WRAP, described the findings as 'shocking'.

'Food waste has a significant environmental impact. This research confirms that it is an issue for us all, whether as consumers, retailers, local or central government. I believe it will spark a major debate about the way food is packaged, sold, stored at home, cooked and then collected when it is thrown out.'

She added:

'What shocked me the most was the cost of our food waste at a time of rising food bills, and generally a tighter pull on our purse strings. It highlights that this is an economic and social issue, as well as about how much we understand the value of our food. Tackling the problem of food waste will be at the heart of WRAP's work over the next three years.'

Julia Falcon of WRAP's campaign, Love Food Hate Waste, said:

'This report shows we could all be saving money and time by making better use of our food. We've found there's a real demand for quick and easy ideas and Love Food Hate Waste can help with tips which turn into good habits in the kitchen.'

Environment Minister Joan Ruddock said:

'These findings are staggering in their own right, but at a time when global food shortages are in the headlines this kind of wastefulness becomes even more shocking.

'This is costing consumers three times over. Not only do they pay hard-earned money for food they don't eat, there is also the cost of dealing with the waste this creates. And there are climate change costs to all of us of growing, processing, packaging, transporting, and refrigerating food that only ends up in the bin.

'Preventing waste in the first place has to remain a top priority. WRAP's advice on the changes everyone can make to ensure they cut their own waste – and their own bills – makes sense all round.'

Notes
The Food We Waste Report:

This study which is believed to be the first of its kind in the world, consisted of a detailed survey of households and a physical analysis of their waste.

A representative sample of 2,715 households in England and Wales was interviewed, and several weeks later, 2,138 of them had their waste collected for analysis – with their signed consent.

The research was designed so that WRAP could quantify the amounts and types of food waste being produced, but also made links between this and the attitudes displayed by, and disposal options available to the household.

Key facts from the Love Food Hate Waste campaign
⇨ In the UK we are throwing away one-third of the food we buy.

That's like one in three bagfuls of food shopping going straight in the bin.
⇨ We throw away 6.7 million tonnes of food each year in the UK, when most of this food could have been eaten. (It's not just peelings and bones –it's good food.) That's equivalent to filling Wembley Stadium with food waste 8 times over!
⇨ In terms of environmental impact – producing, storing and getting the food to our homes uses a lot of energy. The carbon impact of food waste is enormous. Tackling it would provide a carbon benefit equivalent to taking 1 in 5 cars off UK roads.
⇨ Most of the wasted food reaches landfill sites where it emits methane, a powerful greenhouse gas.
⇨ High economic cost – at least £10bn worth of food that could have been eaten is thrown out every year.
⇨ We throw food out for two main reasons: food gets forgotten and is left unused; we serve up too much and don't use leftovers.

For more information on the Love Food Hate Waste campaign, visit: www.lovefoodhatewaste.com
8 May 2008

⇨ The above information is re-printed with kind permission from WRAP. Visit www.wrap.org.uk for more information.

© WRAP

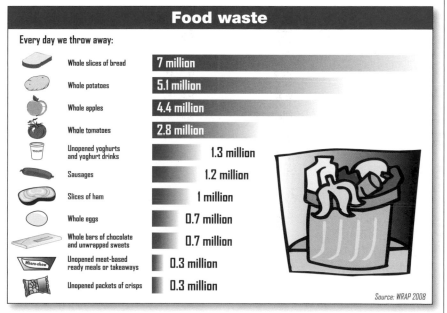

Food waste

Every day we throw away:

Whole slices of bread	7 million
Whole potatoes	5.1 million
Whole apples	4.4 million
Whole tomatoes	2.8 million
Unopened yoghurts and yoghurt drinks	1.3 million
Sausages	1.2 million
Slices of ham	1 million
Whole eggs	0.7 million
Whole bars of chocolate and unwrapped sweets	0.7 million
Unopened meat-based ready meals or takeaways	0.3 million
Unopened packets of crisps	0.3 million

Source: WRAP 2008

Scale of packaging waste problem

Information from the Local Government Association

The Local Government Association has today published its second investigation into the weight of retailer packaging and how much of it can be recycled.

It shows that up to 38 per cent of packaging in a regular household shopping basket cannot be recycled. The first survey, published in October 2007, put that figure at 40 per cent.

The British Market Research Bureau (BMRB) was commissioned by the LGA to buy a range of common food items from eight retailers. Analysis of the purchases found that local retailers and market traders produced less packaging and more that could be recycled than the larger supermarkets.

The supermarket with the heaviest packaging was Lidl (813 grams). Lidl and Marks & Spencer had the lowest level of packaging that could be recycled (62 per cent). Asda's packaging weighed least among the major supermarkets – 646 grams, 69 per cent of which was recyclable. Packaging weighed least of all in a local market – 617 grams – and 76 per cent of it was recyclable.

Most retailers have a lower weight of packaging than in the first survey – the average weight of packaging has reduced by five per cent – though the proportion that is recyclable has changed little.

Cllr Paul Bettison, Chairman of the LGA Environment Board, said:

'The days of the cling film coconut must come to an end. We all have a responsibility to reduce the amount of waste being thrown into landfill, which is damaging the environment and contributing to climate change.

'Families will be pleased to see that more packaging in their shopping baskets can now be recycled. However, this survey shows there is still a lot further to go. Reducing packaging is vital if we are to avoid paying more landfill tax and EU fines, which could lead to cuts in frontline services and increases in council tax.'

Up to 38 per cent of packaging in a regular household shopping basket cannot be recycled

---

Recycling rates have increased to 33 per cent in England as local people do their bit to reduce the amount of waste being thrown into landfill – but this figure must continue to rise. Councils have to pay £32 in tax for every tonne of rubbish that is sent to landfill. That figure will rise to £48 a tonne by 2010. In addition, from 2010 councils face EU fines of £150 for every tonne that is dumped, which could cost an estimated £200 million by 2013.

Cllr Bettison added:

'Some packaging is often needed, particularly to prevent food becoming spoilt and then ending up in landfill sites. Many retailers are also taking some very positive environmental initiatives, such as Marks & Spencer agreeing to work with councils to part-fund a recycling plant in East London. The LGA is in very constructive discussions with some of the leading supermarkets, and we hope this dialogue will lead to further steps in the direction of reducing packaging.'

The LGA is calling on the Government to make retailers and producers responsible for funding the collection of packaging so they are incentivised to cut back on it in the first place. Other countries in the European Union operate a system where companies contribute towards recycling services and household collections.

'Many countries on the continent operate a system where retailers contribute towards household collection and recycling services. This acts as an incentive for them not to produce excessive packaging in the first place. Government urgently needs to change its approach so retailers are incentivised to minimise unnecessary packaging and support maximum recycling.'

29 May 2008

⇨ The above information is reprinted with kind permission from the Local Government Association. Visit www.lga.gov.uk for more information.

Q&A: plastic bags

Plastic bags have become the subject of huge debate in early 2008. Why is everyone suddenly so interested?

Why are plastic bags in the news?

Today, M&S announced that the chain would charge food customers 5p for every plastic carrier bag they use.

The charge is aimed at reducing demand for the bags, and Marks & Spencer says the money raised from the levy will be spent on improving parks and play areas across the country.

Earlier this week, the *Daily Mail* dedicated its first nine pages to the issue of plastic bags.

Accompanying the front-page splash headline Banish the Bags, the *Mail* showed contrasting pictures of a British family carrying numerous plastic bags from their weekly shop and a turtle swimming among discarded plastic bags.

The paper was aiming to highlight the 13bn bags given away each year by British retailers and graphically depicted the environmental impact of plastic waste in a series of images showing animals chomping on bags or swathed in plastic.

And on Tuesday, China announced that its war against 'white pollution' had claimed its first large-scale victim with the closure of the country's biggest plastic bag manufacturer.

The shutdown of Suiping Huaqiang Plastic, which employs 20,000 people, highlighted the social costs of a government drive to clean up one of the world's most polluted environments.

It comes less than two months after the state banned production of ultra-thin bags and ordered supermarkets to stop giving away free carriers from June 1.

What are we doing about the problem of plastic bags in the UK?

London is attempting to ban the millions of disposable shopping bags given away by shops each year.

A bill is currently being read in the Commons that was entered by the chief executives of the 33 London

By Jessica Aldred

councils who received overwhelming public support for the proposals, which were first put forward in July 2007.

More than 90% of organisations said they wanted a complete ban, or would support a 10-15p levy on every bag.

The government has so far resisted a national ban or a levy, preferring a voluntary agreement with supermarkets to reduce the 'overall environmental impact' of carrier bags by 25% by the end of 2008.

A statement on the Department for Environment, Food and Rural Affairs' website says: 'There is no clear evidence that such a tax would be beneficial on either broad environmental or litter grounds. This is because people would be encouraged to use bags made from other materials or alternative forms of packaging, which may be equally or more damaging to the environment.'

Which other cities in the UK have banned the plastic bag?

Traders in 80 mainly small towns round Britain have either introduced a voluntary ban or are considering one as a way to reduce landfill.

Brighton and Hove council is the largest authority in Britain to offer support for a voluntary ban.

Cities around the world, from San Francisco to Dacca in Bangladesh, have vastly reduced the number of bags being issued by imposing taxes.

Many towns in Britain have been inspired by the action of Rebecca Hosking, who persuaded all 34 local shops in her home town of Modbury in south Devon to substitute their plastic bags with reusable cloth bags.

The BBC camerawoman was moved to do this when she saw albatrosses, turtles and dolphins choking to death on plastic while filming in the Pacific last year.

Why are plastic bags so bad?

Around 13bn plastic bags are given free to UK shoppers every year.

The bags can take between 400-1,000 years to break down, and like all forms of plastic they do not biodegrade. Instead they photodegrade, breaking down into smaller and smaller toxic bits that contaminate soil, waterways and oceans, entering the food chain when ingested by animals.

Many plastic bags end up as waste on our beaches, streets and parks. When a plastic bag enters the ocean it becomes a harmful piece of litter. Many marine animals mistake plastic bags for food and swallow them, with painful and often fatal consequences. Nearly 90% of floating marine litter is plastic.

What are retailers doing?

Marks & Spencer announced its 5p charge plans for England last year following a successful trial in Northern Ireland, which led to a 66% reduction in the number of bags used by customers.

Sainsbury's reported a fall in the use of free carrier bags of 10% during six months last year, while the use of reusable bags rose by nearly 50%. Since February 2007, all of its free bags have been made from 33% recycled plastic. A 'bag for life' cost 10p, though they are sometimes given away. The supermarket said that if all its customers reused these bags 20 times, it would save 90m disposable bags a year. It has also sold £5 cotton bags designed by Anya Hindmarch with the slogan 'I'm not a plastic bag', and had a one-day moratorium on plastic bags in April 2007.

Tesco, the UK's largest supermarket, gives out 4bn free plastic bags each year, but allocates reward points to shoppers who refuse them. Online shoppers can choose bag-free delivery. The supermarket's carrier bags are biodegradable.

Asda gives out free disposable bags, although all those returned to the store are recycled. The supermarket sells reusable bags for 5p and jute bags for 97p. Cash register operators are instructed to offer fewer plastic bags to customers.

Waitrose was the first supermarket to introduce reusable bags, which it sells for 10p. It claims that in 2005 this helped reduce the number of disposable bags distributed by 54m. But it still hands out 250m free disposable bags a year to its customers.

What can I do?

Refuse plastic bags in shops, and try to recycle or reuse the ones you do have. Buy a woven cotton 'eco-shopper' bag and keep it with you for when you need to go to the shops.

Can I get my town to ban plastic bags?

Yes, but you must start the campaign yourself, or with a group. Don't rely on councils or supermarkets. Get the trust of the traders by approaching them directly – a handout is not enough.

Gauge public support to encourage supermarkets and multiples to take part, and learn about what plastics are doing to the environment and research every type of alternative bag on the market.

Set a date for the ban and go for it.

28 February 2008

© *Guardian Newspapers Limited 2008*

Carrier bags

Information from Defra

What's the problem?

The 13 billion carrier bags which are distributed in the UK each year comprise a comparatively small part of the domestic waste stream. However, they can be a particularly visible form of litter when discarded irresponsibly, and their disposable, easily-substituted nature make them a symptom of our 'throwaway society' – and public opinion recognises this. Currently, each adult receives on average nearly 300 disposable bags every year.

The Government's Waste Strategy for England 2007 set out the Government's aim that free, single-use carrier bags (both paper and plastic) should become a thing of the past. Behaviour change of this kind is vital if we are to achieve a more sustainable lifestyle.

What have we done so far – voluntary agreement

On 28 February 2007, a voluntary agreement was announced with UK retailers to reduce the overall environmental impact of carrier bags by 25% by the end of 2008. 22 major retailers and six trade associations signed up to the agreement.

This agreement covers all types of carrier bags, including both paper and plastic. It gives retailers from across the sector the flexibility to respond in a way which is best suited to their business and the needs of their customers. At the same time it encourages individuals to play their part in minimising environmental impacts and to change their shopping habits to reuse and recycle more.

We welcome the action taken by retailers to cut the environmental impacts of their carrier bags, and especially the leadership shown by several retailers who have taken steps to secure greater changes in public behaviour. However, it is clear that the current voluntary arrangement has not resulted in a substantial reduction in the number of bags distributed, and will not get us close to achieving our goal of phasing out free, single-use carrier bags.

Budget 2008: charges for single-use carrier bags

The Government has, therefore, decided that stronger action needs to be taken. It was announced in the Budget, on 12 March 2008, that the Government will bring forward legislation in the Climate Change Bill to enable us to require retailers to impose a minimum charge on single-use carrier bags, if sufficient progress is not made on a voluntary basis. These powers will come into force in 2009. The Government will consult in the meantime on the operation of the charge and how to ensure that any money raised goes to environmental charities. See the HM Treasury website for Budget announcements.

Key facts and figures

⇨ More than 13 billion carrier bags are distributed in the UK every year. These account for approximately:

↪ 0.3% of the domestic waste stream (HM Treasury's Plastic Bags Tax Assessment in December 2002);

↪ 0.1 – 1% of visible litter in the UK (Industry Council for Packaging and the Environment [INCPEN] Visible Litter Study 2004);

↪ 2% of total litter on UK beaches (Marine Conservation Society [MCS] annual beach litter survey 2004).

⇨ Limiting the use of disposable bags is an important way in which each of us can take action, as each adult receives on average nearly 300 every year.

⇨ 88% of shoppers currently put all their shopping into free carrier bags. On average shoppers take 3-4 bags at every shopping trip.

⇨ 45% of shoppers claim to have bought a Bag for Life but only 12% use one regularly.

12 March 2008

⇨ The above information is reprinted with kind permission from Defra. For more information on this and other topics, please visit www.defra.gov.uk

© *Crown copyright*

Litter

Information from ENCAMS

Is it a crime to drop litter?
Yes, if it happens in a public place. The average fine is around £95 although a court does have the power to fine someone up to £2,500 (a Level 4 offence on the Standard Scale). Cases are heard in the Magistrates' Court. Approximately 400 people were prosecuted last year by the police for littering. Alternatively, in some areas you could get a £50 fixed penalty fine for littering from the local authority 'litter warden'.

What is litter?
Legally speaking, the word litter is given a wide interpretation. Litter can be as small as a sweet wrapper, large as a bag of rubbish or it can mean lots of items scattered about. ENCAMS describes litter as 'Waste in the wrong place caused by human agency.' In other words, people make litter. The Offence of Leaving Litter (section 87 of the Environmental Protection Act 1990) says that if a person drops, throws, deposits or leaves anything so as to cause defacement in a public place, they could be committing a littering offence. So always put rubbish in a bin, or take it home.

What can I do if I see someone drop litter?
If you are with someone you know and they drop litter you may feel safe telling them they shouldn't. However, even though it is infuriating to see someone littering, ENCAMS does not want you to put your personal safety at risk. The same applies if you see litter thrown from cars. Police officers or litter wardens are empowered and trained to deal with offenders. If you have information about a littering incident you could report it to the police, the local authority or a litter warden, but it is up to them to decide whether they should wish to proceed any further. Whilst it is possible to take a private prosecution, it would be at a person's own expense and you will need strong evidence to prove your case in court.

Who should clear away litter?
Your local authority has a legal duty (so far as practicable) to clear litter and refuse from public places for which it is responsible such as streets, parks, playgrounds, tourist beaches and pedestrianised areas. If a piece of private land is littered, the owner should accept responsibility for clearing the litter. But we should all help by not dripping litter in the first place.

Who should I report it to at the council if my street needs cleaning?
Each local authority will have its own title for the department that carries out street cleaning. Usually it will be called something like Cleansing or Environmental Services or Environmental Health. When you telephone state that your complaint is about street cleansing (or refuse collection or fly tipping removal), ask which department is responsible then ask to be put through to them. Find out what action they intend to take and when. Note the date, time and the name of the person you speak to for future reference.

⇨ The above information is reprinted with kind permission from ENCAMS. Visit www.encams.org for more.

© ENCAMS

The problem with litter

Information from the Campaign to Protect Rural England

Why we need to stop litter and fly-tipping
⇨ It costs taxpayers in excess of half a billion pounds annually to clear the streets of England, and that doesn't include parks or other public spaces.
⇨ Nearly half the population (48%) admit to dropping litter.
⇨ The amount of litter dropped yearly in the UK has increased by 500% since the 1960s.
⇨ MPs get more letters complaining about litter and dog fouling than anything else.

Campaign to Protect Rural England

⇨ It is illegal to drop litter, and you can be fined up to £80 on the spot if you're caught littering.
⇨ According to a MORI poll in 2002, clean streets come only second to crime and hospitals in a survey of local issues.

⇨ A MORI poll in July 2007 found the public more concerned about litter and graffiti than they were about climate change.
⇨ Nationally, seven out of ten items of litter are food related.
⇨ An estimated 122 tons of cigarette butts and cigarette-related litter is dropped every day across the UK.
⇨ 1.3 million pieces of rubbish are dropped on Highways Agency roads alone every weekend (over a year this adds up to a whopping 67.2 million pieces of rubbish).

A third of drivers admitting to throwing litter while on the road.

⇨ The rat population has boomed to 60 million due to the huge amounts of litter around. This means there are now almost as many rats as people in the UK.

⇨ Over 69,000 animals were killed or injured by litter last year in Britain.

⇨ Litter, such as cigarette butts, plastic bags and other plastics, harms animal and marine life in a variety of ways such as water pollution, when these items are mistaken for food and when creatures get caught up in plastics and get strangled. It is estimated that every year over 1 million seabirds and 100,000 turtles and sea mammals die of litter-related causes.

⇨ Over 373,000 pieces of litter were found on beaches in the UK in 2006 – equivalent to 1,989 items per kilometre – the Marine Conservation Society reports a 90% increase in beach litter since 2004.

⇨ The general level of litter has dropped from 'satisfactory' to 'unsatisfactory' in the last 12 months by Government's own measures and standards.

⇨ The most recent Government study on local environmental cleanliness showed that there has been a significant increase in roadside litter to moving it into the 'unsatisfactory' category for the first time in the six years the survey has been conducted.

⇨ Litter makes an area look dirty and uncared for and attracts more litter. Littered areas are not pleasant to be in and are less likely to be used by people. In contrast, people are more reluctant to litter clean areas.

⇨ Littered items are a lost resource. When things that could otherwise be recycled, like PET, glass bottles and paper, are littered, they do not end up in the recycling stream.

Biodegradability

Any type of litter takes a long time to disappear naturally, so whatever the material the right thing to do is not to drop it in the first place. Degradability depends on climate and circumstances, but under unfavourable conditions estimated time spans are:

⇨ orange peel and banana skins – up to 2 years;

⇨ cigarette butts – up to 2 years;

⇨ plastic bags – 10-20 years;

⇨ tin cans – 50 years;

⇨ aluminium cans – 80-100 years;

⇨ plastic bottles – indefinitely;

⇨ glass – indefinitely.

Facts about fly-tipping

⇨ Fly-tipping is the common term used to describe waste illegally deposited on land and in simple terms, a single bin bag upwards to thousands of tonnes of construction and demolition waste may constitute a fly-tip.

⇨ The illegal disposal of waste is an anti-social behaviour that is adversely affecting the amenity of our local environments and reducing civic pride. Fly-tipping poses a threat to humans and wildlife, damages our environment, and spoils our enjoyment of our towns and countryside.

⇨ A new incident of fly-tipping occurs every 12 seconds at a cost to the public purse of around £72 a minute.

⇨ £73.7 million – the estimated cost of clearance of illegally dumped waste reported by local authorities in 2006/07.

⇨ £47 million – the estimated cost of clearing fly-tipping from agricultural land alone in 2005/06 (Environment Agency). Areas subject to repeated fly-tipping may suffer declining property prices and local businesses may suffer as people stay away.

⇨ Local authorities in England reported that they had dealt with more than 2.6 million incidents of fly-tipping in 2006/07 – up five per cent on 2005/06.

⇨ There were only 1,796 successful prosecutions of fly-tipping in 2006/07 – a 1 in 1,460 chance of being successfully prosecuted.

⇨ Over half (56%) of fly-tipping incidents reported were in alleyways.

⇨ Black bags full of domestic rubbish account for 63% of all fly-tipping.

⇨ 77% of fly-tips involved household waste – a 5.4 per cent increase on 2005/06.

⇨ 95% of farmers have cleared up other people's rubbish from their land.

⇨ Most people fly-tip to avoid paying the disposal fee called the landfill tax. Household rubbish is already paid for through council tax, but other waste is not, and a charge generally exists to have this waste removed or even to drop the waste off at a licensed tip yourself.

⇨ Fly-tipping can incur fines of up to £20,000 and/or 6 months' imprisonment. Fines are unlimited if the case goes to the Crown Court or up to 2 years' imprisonment, and up to 5 years if hazardous waste is dumped.

⇨ The above information is reprinted with kind permission from the Campaign to Protect Rural England. Visit www.cpre.org.uk for more information.

© CPRE

WHAT DO YOU THINK WILL LAST LONGER, THE HUMAN RACE OR OUR RUBBISH?

British waste adds to crisis across China

⇨ *One-fifth of rubbish in province is imported.*
⇨ *Recycling firms relocate to get round crackdown.*

By Jonathan Watts

British high street waste is fouling streams and ditches in China despite promises of an environmental crackdown by both governments.

Mai village in Guangdong province, southern China, suffers from a Made-in-Britain eyesore: Tesco and Argos plastic bags choke the waterways, snag on tree branches and contribute to a rotting stench during floods and hot weather. There is even a green and white Help The Aged carrier bag printed with a slogan proclaiming the charity's fight against 'poverty, isolation and neglect'.

It is a side-effect of globalisation. Many of these products were manufactured in China, shipped to the UK for use and sent 5,000 miles back for disposal.

China exported £12.6bn worth of manufactured goods to the UK last year and received an estimated 1.9m tonnes of rubbish in return. Under EU regulations member countries are not allowed to dump garbage overseas, but are permitted to send sorted waste for recycling.

Environmentalists say this is irresponsible because much of the recycling is carried out in poorly-regulated communities, where health risks and pollution worries are a low priority.

Guangdong is scattered with scavenging centres. In Guiyu and Qingyuan small family-run businesses chop up and melt down toxic plastics and metals from discarded computers, printers and mobile phones. In Nanhai and Shunde factories deal with mounds of plastic bags and bottles. About 20% of the waste comes from overseas according to local sources.

A series of exposés in the domestic and foreign media prompted the government to crack down on the business earlier this year. Guangdong's provincial government banned unlicensed businesses and individuals from importing plastic waste and suspended operations at factories that failed to meet environmental standards.

Last month factories in the most notorious district, the Lianjiao area of Nanhai, were shut down. But most firms simply relocated.

Two hours' drive away a new recycling centre is under construction in Shijing village, which is now littered with scrapheaps. The dealers said they would no longer touch foreign waste.

In nearby Shenzhen and Shunde businessmen were still reprocessing carrier bags and other UK waste from the UK. 'It can be done as long as the plastic is well enough packaged to get through customs,' said the owner of one factory.

At 'plastic street' in Mai village, dozens of small plastic recycling firms line the road. Most of the work is done by migrant workers who are paid about £50 a month.

Thousands of plastic carrier bags were being blown into ditches and waterways, creating an eyesore and a bad smell. Students at the local school said the stench came into their classrooms and got worse when the fetid stream floods.

The sanitary department of Shunde township said it was unaware of the mess in Mai village.

'We have a project to clean up villages in this area but we haven't got round to Mai yet,' said a spokesman. The provincial government declined to comment.

Britain supports the recycling business. 'It allows for a more sustainable use of world resources, but it should be carried out under strict environmental controls,' said the UK consulate in Guangzhou. When told of the impact on Mai village it said individual companies should take more responsibility.

Greenpeace believes that wealthier countries should deal with their waste problems at home, rather than exporting them to developing countries, which have to pay the environmental costs.

'If we can stop the waste trade I am sure it will lead to more sustainable development around the world,' said Kevin May, toxics campaign manager at Greenpeace's office in Beijing.
31 March 2007
© *Guardian Newspapers Limited 2008*

Waste at home

Information from Waste Online

What we produce!

We all produce waste of some sort, whether it is the empty drinks can, or the grass clippings from the garden. We estimate that nearly 30million tonnes of household waste were collected in the UK in 2003/04. That's over 500 kg, or half a tonne, of rubbish per person per year!

So where does it all go?

A total of:

⇨ 72% of municipal waste is landfilled – which means it's buried in the ground.

⇨ 9% is incinerated – which means it's burnt – this is also called energy from waste.

Dealing with our rubbish in this way is not an ideal solution. When we bury or burn our rubbish we are losing valuable natural resources and wasting the energy, water and transport costs used in its production. Landfilling and incineration can harm the environment if not properly managed. Many landfill sites are nearly full and we are rapidly running out of suitable land, close to where the rubbish is produced, for new sites. In any case, these sites are often unwelcome neighbours – we keep producing the rubbish, but we don't want it disposed of near to where we live.

The alternative?

Reduce, Reuse, Recycle

We would all benefit from:

⇨ Reducing the amount of rubbish we create.

⇨ Reusing as much of our rubbish as possible.

⇨ Increasing the 19% of waste that we currently recycle and compost (although the latest figures suggest we are now recycling 23%).

That is what the 3Rs are about – reduce, reuse, recycle.

Reducing

Every year the amount of rubbish we produce increases and this leads to increased costs for society – both financial and environmental. The majority of the resources that we use to make things – only to throw them away – can't be replaced. Throwing away our rubbish puts pressure on the environment – not only from the landfills and incinerators, but also because we have to extract and process even more resources, and transport our new goods and our old rubbish so increasing vehicle emissions. As consumers, we have the ability to reverse this trend – buy only the right quantity of what we really need, choose products with less packaging, and buy from producers employing sustainable practices.

Reusing

We can cut down on the amount of rubbish we have to get rid of by reusing our materials. Computers, furniture, clothing – so many items can be reused. Setting the printer to print on both sides of a sheet of paper, repairing our broken appliances and shoes or finding a charity that will make use of them – we help ourselves and others, and delay the point at which materials become waste.

Recycling

Putting materials aside for recycling helps in many ways: we send less rubbish to landfill or incineration, and we save valuable materials and energy – for example, plastic bottles can be converted into fleeces and garden furniture, whilst recycling aluminium cans saves 95% of the energy used in making a new can. New technologies are furthering our ability to recycle what was previously our waste and turn it back into the resources that we need.

If you would like to recycle more, but do not know where to take your recyclables, then visit the site www.recyclenow.com. Here you will also find contact details for your local authority.

Buy recycled

If you find that a material is not being recycled in your area, it may be because the markets are not strong enough for the local authority to pay for collecting it. To help alleviate this, support those industries that use recycled materials by buying recycled products. Buying recycled 'closes the loop' in recycling – remember that it's not enough just to recycle, buying recycled ensures that the materials you send for recycling are actually used again.

Many landfill sites are nearly full and we are rapidly running out of suitable land

---

You may also be interested in other ways to use your power as a consumer to help promote sustainability – visit green choices for further suggestions on environmentally friendly shopping.

Sustainable solutions

The problem of what to do with the waste we produce is worldwide though the solutions have to be provided locally. In the UK there are many organisations – governmental and independent, local and national who are trying to move Britain forward towards a sustainable society.

The reports and information they have produced are a valuable resource to help us all move on and really begin the 'good riddance of bad rubbish'.

9 June 2008

⇨ The above information is reprinted with kind permission from Waste Online. Visit www.wasteonline.org.uk for more information.

© *Waste Online*

Tips to reduce waste

Information from Energy Saving Trust and WRAP

Recycling

Recycling reminders

Making a note on your calendar or fridge will provide a visual reminder of the day/dates your recycling is collected. Nine out of ten UK residents now have access to a doorstep recycling collection.

Whole house recycling

It's common for people to recycle in the kitchen but forget other rooms. Recycle the waste from all the bins in your house. For example, if you can recycle plastic bottles then all plastic bottles can go into your collection, including shampoo bottles.

Keep it simple

Keep your recycling bin next to your main bin so you can take your rubbish and recycling out at the same time. If you don't already have a recycling box or bag you should contact your local council. See the Recycle Now website at www.recyclenow.com for more information about recycling.

Nine out of ten UK residents now have access to a doorstep recycling collection

Wasted food

Long-life food

A third of the food we buy in the UK ends up being thrown away so keeping your fridge, freezer and cupboard stocked with long-shelf-life basics means you'll always have ingredients at hand to make the most of fresh food. There are recipes and practical ways to make the most of the food you've bought at the Love Food Hate Waste website (www. lovefoodhatewaste.com).

Reusing leftovers

Wasted food is a waste of money and, when sent to landfill, a major contributor to climate change because it breaks down to produce methane, which is a powerful

greenhouse gas. If we in the UK stopped wasting food that could have been eaten, for example by reusing leftovers rather than throwing them away, it would have the same impact on carbon emissions as taking one in five cars off our roads.

Home composting

Mix it up

By composting at home you can save as much carbon as your kettle produces annually. The key to good compost is to get a good mix of wet and sappy materials with dry and fibrous ones. This will ensure your bin has all the moisture and air it needs to compost successfully. Visit the WRAP home composting website to find out more (www.recyclenow.com/home_composting).

Keep a kitchen compost caddy

Over 30% of an average household bin can be composted at home, from vegetable peelings and teabags, to egg boxes and shredded paper. Using a container or kitchen caddy to collect your compostable waste from the house will save you having to make a trip to the compost bin every day.

Other

Reduce unwanted mail

Many organisations now offer secure

online billing rather than sending a paper bill each month. Eliminate junk mail by registering for free with the Mail Preference Service to have your name, as well as those of previous residents, removed from direct mail lists.

Ditch the disposables

Use products with a longer life, such as energy-saving light bulbs, which can last around ten times longer than standard bulbs. You could also buy more concentrated versions of many products such as your washing detergent or washing-up liquid.

Reuse where you can

Over 13 billion plastic bags are distributed in the UK every year. That's around 215 per person for the population of the UK. Reuse your plastic bags as often as you can or buy a stronger canvas or reusable shopping bag. Donate unwanted items such as clothes, books, CDs and furniture to charity shops or join a gift community such as Freecycle.

⇨ Reprinted with kind permission from the Energy Saving Trust and the Waste Resources Action Programme (WRAP). Visit their websites at www. energysavingtrust.org.uk and www. wrap.org.uk for more.

© Energy Saving Trust/WRAP

A strategy to cut waste

Information from Directgov

The Environment Secretary David Miliband today published a new strategy for cutting waste, and said that everyone – businesses, individuals, local authorities and the government – has a role to play by reducing the waste they produce. He said that this would be an essential part of the drive to tackle climate change – landfilled waste is a major source of methane, a potent greenhouse gas, while reducing and recycling waste saves energy and raw materials.

Following calls from local authorities, the government is also launching a parallel public consultation on removing the ban on local authorities introducing financial incentives for recycling. Any such schemes will have to return all revenues back to local residents.

The main points of the waste strategy include:

⇨ more effective incentives for individuals and businesses to recycle waste, leading to at least 40 per cent of household waste recycled or composted by 2010, rising to 50 per cent by 2020. This is a significant increase on the targets in the previous waste strategy, published in 2000.

⇨ a greater responsibility on businesses for the environmental impact of their products and operations through, for example, a drive to minimise packaging and higher targets for recycling packaging.

⇨ a strong emphasis on waste prevention with householders reducing their waste (for example, through home composting and reducing food waste) and business helping consumers, for example, with less packaging. There will also be a new national target to help measure this – to reduce the amount of household waste not reused, recycled or composted from 22.2 million tonnes in 2000 to 12.2 million tonnes by 2020 – a reduction of 45 per cent.

⇨ the government has agreed with the Direct Marketing Association to develop a service so that people will be able to opt out of receiving unaddressed as well as addressed direct mail. The government is also considering moving towards an approach where people would only get direct mail if they opted in by placing their name on the direct mail register.

⇨ working with retailers for the end of free single-use bags. This could involve retailers only selling long-life bags, or retailers charging for disposable bags and using the proceeds to sell long-life bags at a discount.

⇨ a challenge to see recycling extended from the home and office and taken into public areas like shopping malls, train stations and cinema multiplexes, so that it becomes a natural part of everyday life. To help deliver this, the government is working with owners of public spaces to draw up guidance and a voluntary code of practice to be published by the end of the year. Groups including the Airport Operators Association, British Council of Shopping Centres, Earls Court and Olympia Group, Highways Agency and the Local Government Association have already come out in support of this. In particular, the Royal Parks has committed to putting recycling bins in all its parks within the next 12 months and the Association of Event Venues says its members plan to install recycling bins for waste brought in by audiences at major events.

⇨ subject to further analysis and consultation, banning biodegradable and recyclable waste from being put into landfill sites.

⇨ an increase in the landfill tax escalator by £8 per year from 2008 until at least 2010/11 – announced by the Chancellor in March. Partly as a result of this, business waste landfilled is expected to fall by 20 per cent by 2010 compared with 2004.

⇨ increasing the amount of energy produced by a variety of energy from waste schemes, using waste that can't be reused or recycled. It is expected that from 2020 a quarter of municipal waste – waste collected by local authorities, mainly from households – will produce energy, compared to 10 per cent today.

Everyone – businesses, individuals, local authorities and the government – has a role to play by reducing the waste they produce

---

David Miliband said:

'We need to not only recycle and reuse waste, but also prevent it in the first place. And there's a particular challenge for businesses to produce less waste with their products, so consumers have less of it to dispose of.

'The result will be a win for individuals, who will have a cleaner, safer local environment, while potentially saving money, and a win for the wider environment because it'll reduce landfilled waste which contributes to climate change.

'This strategy sets out how we can achieve this. It provides a range of tools for local authorities, businesses and individuals to do the job. It calls for action from all, without imposing one-size-fits-all solutions. It empowers local authorities to make the right decisions for local circumstances in consultation with their local populations.'
24 May 2007

⇨ The above information is reprinted with kind permission from Directgov. Visit www.direct.gov.uk for more information.

New powers needed to tackle litter louts

Information from the Local Government Association

Up to 70 per cent of all litter offences go unpunished because of a legal loophole that makes it almost impossible to tackle louts who throw rubbish from cars, council leaders warned today.

Local authorities across the country have complained that the law is inconsistent and inadequate, and that enforcement loopholes leave them powerless to take action against offenders.

Simple revisions to existing legislation, which would place responsibility for litter dropped from a vehicle on the registered keeper, would give councils the enforcement powers they need.

Rubbish thrown from vehicles is estimated by some councils to account for 70 per cent of street litter in some areas – but local authorities trying to tackle the problem are finding themselves entangled in a legal Catch-22 situation.

Cllr Paul Bettison, Chairman of the Local Government Association Environment Board, said:

'At a time when councils are coming under increasing pressure to deal with littering, some of the current legislation is a mire of confusion. Registered keepers of vehicles can be prosecuted for speeding unless details of the offender are given, so why not for littering? As the law stands, local authorities are effectively charged with tackling this anti-social behaviour with their hands tied behind their backs.

'Councils strive to keep the places where people live clean and tidy. Litter is environmental vandalism – it's unpleasant, unnecessary and unacceptable. Simple changes to the existing law would close the loopholes currently causing such confusion and inconsistency.'

Under current DEFRA guidance, councils are advised not to issue fixed penalties for littering from vehicles unless the offender can be positively identified (in accordance with Section 87 of the Environmental Protection Act 1990). The reality for environmental officers is that positively identifying a person in a moving vehicle is nigh-on impossible – and this leads to inconsistencies in enforcement.

Some local authorities do not issue any fixed penalties, while others issue penalties to the registered keeper of the identified vehicle. But if the penalties are not paid, authorities are faced with prosecuting the registered keeper unless the offender's details are voluntarily provided – which presents onerous legal challenges for the enforcing authority.
14 April 2008

⇨ The above information is reprinted with kind permission from the Local Government Association. Visit www. lga.gov.uk for more information.
© Crown copyright

Government wants us to recycle on the go

By Sam Bond

Apparently happy with progress on the home front, Government now hopes to encourage us to embrace recycling while we're out and about.

Launching the Recycle on the Go pilot scheme in London's Hyde Park on Monday, environment Minister Joan Ruddock told edie that the idea was to provide recycling bins in convenient locations to give people the opportunity to do the right thing.

'We're saying, "look, we know you're not going to take those cans and bottles home with you but now you can recycle them rather than just throwing them in the bin",' she said.

'We know people are prepared to recycle and this is about giving them the opportunity to do so.'

She said that previously litter collected in park bins would go to landfill as it was inefficient to try to sort it for recycling.

'Normal bin waste is usually highly contaminated so it is difficult to recycle,' she said.

She acknowledged there was likely to be a relatively high level of contamination initially with these bins, as there had been with kerbside recycling schemes when they were first rolled out, but she said people had learned how to sort their waste at home and she was confident they could do so while out and about too.

She accepted that two bins in one park were unlikely to have a significant impact on general recycling rates, but argued that it was better to start with a pilot scheme to assess the public appetite for recycling while they're on the move, before rolling out a wider initiative.
3 June 2008

⇨ The above information is reprinted with kind permission from edie. Visit www.edie.net for more information.
© edie

Recycling facts and figures

Information from Recycling Guide

UK households produced 30.5 million tonnes of waste in 2003/04, of which 17% was collected for recycling (source: defra.gov.uk). This figure is still quite low compared to some of our neighbouring EU countries, some recycling over 50% of their waste. There is still a great deal of waste which could be recycled that ends up in landfill sites which is harmful to the environment.

Recycling is an excellent way of saving energy and conserving the environment. Did you know that:

⇨ 1 recycled tin can would save enough energy to power a television for 3 hours.

⇨ 1 recycled glass bottle would save enough energy to power a computer for 25 minutes.

⇨ 1 recycled plastic bottle would save enough energy to power a 60-watt light bulb for 3 hours.

⇨ 70% less energy is required to recycle paper compared with making it from raw materials.

Some interesting facts

⇨ Up to 60% of the rubbish that ends up in the dustbin could be recycled.

⇨ The unreleased energy contained in the average dustbin each year could power a television for 5,000 hours.

⇨ The largest lake in Britain could be filled with rubbish from the UK in 8 months.

⇨ On average, 16% of the money you spend on a product pays for the packaging, which ultimately ends up as rubbish.

⇨ As much as 50% of waste in the average dustbin could be composted.

⇨ Up to 80% of a vehicle can be recycled.

⇨ 9 out of 10 people would recycle more if it were made easier.

Aluminium

⇨ 24 million tonnes of aluminium is produced annually, 51,000 tonnes of which ends up as packaging in the UK.

⇨ If all cans in the UK were recycled, we would need 14 million fewer dustbins.

⇨ £36,000,000 worth of aluminium is thrown away each year.

⇨ Aluminium cans can be recycled and ready to use in just 6 weeks.

Glass

⇨ Each UK family uses an average of 500 glass bottles and jars annually.

⇨ The largest glass furnace produces over 1 million glass bottles and jars per day.

⇨ Glass is 100% recyclable and can be used again and again.

⇨ Glass that is thrown away and ends up in landfills will never decompose.

Paper

⇨ Recycled paper produces 73% less air pollution than if it was made from raw materials.

⇨ 12.5 million tonnes of paper and cardboard are used annually in the UK.

⇨ The average person in the UK gets through 38kg of newspapers per year.

⇨ It takes 24 trees to make 1 ton of newspaper.

Plastic

⇨ 275,000 tonnes of plastic are used each year in the UK, that's about 15 million bottles per day.

⇨ Most families throw away about 40kg of plastic per year, which could otherwise be recycled.

⇨ The use of plastic in Western Europe is growing about 4% each year.

⇨ Plastic can take up to 500 years to decompose.

⇨ The above information is re-printed with kind permission from Recycling Guide. Visit www.recycling-guide.org.uk for more information.
© Recycling Guide

Recycled household materials

Materials collected from household sources for recycling: 2006/07, England

- Plastics 0.6%
- Cans 1%
- Textiles 1.3%
- Scrap metals and white goods 7.5%
- Co-mingled 15.4%
- Other 8.9%
- Paper and card 19%
- Glass 10.4%
- Compost 35.9%

Source: Defra. Crown copyright.

Recycling tips

Information from Recycle More

This section gives general advice on rubbish disposal and recycling. Always use your common sense and never place rubbish where it could cause harm. If you are at all unsure contact your Local Authority and speak to the recycling officer.

Asbestos

⇨ A building and insulation material commonly used before the 1970s
⇨ the three main types of asbestos are white, blue and brown
⇨ can only cause harm if the fibres are inhaled
⇨ removal from buildings may disturb the fibres.

Advice: Contact your local council before removal.

Batteries

⇨ Some councils and garages provide facilities for recycling rechargeable batteries and lead acid car batteries
⇨ rechargeable batteries should be returned to the manufacturer where possible for disposal
⇨ there are currently kerbside collection trials in the UK for domestic batteries, find out whether your council is involved.

Advice: Buy rechargeable batteries, or appliances which use mains electricity.

Binoculars

⇨ You can donate your old or unwanted pair of binoculars, telescope, spotting scope or tripod that are in good working order to the RSPB where they will be used for conservational or educational projects.

Advice: Visit www.rspb.org.uk for further information and for the address to send your items to.

Building rubbish

⇨ Includes: bricks, asphalt, glass, metals, plastics, soil, and wood
⇨ most of this waste is created by the construction industry
⇨ architectural salvage yards take some items for resale, and old bricks and timber can often be reused.

Advice: Contact the Association for environmentally conscious building for advice.

Cars and vehicles (ELVs)

⇨ Over 1.8 million tonnes of old vehicles are thrown away in the UK each year
⇨ on average 75% of a vehicle is currently recycled
⇨ scrap merchants are able to recycle old vehicles

Over 1.8 million tonnes of old vehicles are thrown away each year

---

⇨ fly tipping is an offence and abandoned vehicle owners can be traced through the DVLA!

Advice: Abandoned vehicles can be reported to your local council, who can also provide advice on disposal.

Cartons (Tetra Pak)

⇨ Tetra Pak supplies the majority of drinks cartons within the UK
⇨ items made from Tetra Pak include milk cartons, fruit juices, liquid foods such as pasta sauces and some ice cream cartons, look out for the Tetra Pak label
⇨ carton recycling facilities are now available in over 300 Local Authorities within the UK.

Advice: Find out from your local council whether there are plans to introduce Tetra Pak recycling in your area and help us to recycle more!

CDs and DVDs

⇨ You can post your old CDs, DVDs, CDRs, VHS and cassette tapes to The Laundry for recycling
⇨ Polymer Reprocessors recycle all CDs and also the packaging around the CDs
⇨ if you are a business Poly C. Reclaimers will collect your CDs free of charge and plastic waste
⇨ you can also reuse you CDs to make funky drinks coasters
⇨ CDs can also be used to make bird scarers or why not buy a clock kit and create a stylish CD clock!

Advice: Reuse your old CDs and DVDs by taking them to a charity shop. Or you can auction your CDs on "Only One Pound", you can opt to donate 50% or 100% to 2 charities.

Chemicals, paint and oils

⇨ Chemicals are used every day in the home and garden, see the National Household Hazardous Waste Forum
⇨ DO NOT pour chemicals and oil down drains. They can pollute rivers
⇨ used engine oil can be recycled at most civic amenity sites
⇨ old or leftover paints can be used by community groups such as Community RePaint
⇨ plastic bottles which have contained household cleaners can also be recycled (check instructions on the bottle).

Advice: Use environmentally friendly chemicals, most DIY stores stock them. Buy in bulk to reduce packaging.

Electronics

⇨ Some retailers take back old electrical items when delivering a new one
⇨ if your item still works safely, you could sell it. See Yellow pages for second-hand electrical shops
⇨ some charity shops will accept small electrical items
⇨ there are now lots of reuse networks where you can offer items for others to reuse
⇨ waste electronics can sometimes be recycled at your Household Waste Recycling Centre
⇨ mobile phones can be recycled through phone retailers and charities
⇨ If your old iPod is broken and no longer under warranty you can have it repaired or sell it to: www.ukipodrepairs.com

Advice: Try to repair broken items rather than throw them away. Buy durable items with long life cycles.

Furniture

⇨ Local charity shops, schools, and community groups can sometimes use unwanted items

⇨ please note that unwanted sofas and chairs must have the kitc mark to prove they meet British safety standards

⇨ most organisations will not take old beds for hygiene reasons

⇨ make sure all furniture is clean and in good repair before you donate it.

Advice: Old furniture may be very useful to someone else! Donate unwanted items where possible.

Glass

⇨ Bottles and jars are usually separated by colour: brown, clear, and green

⇨ place in the correct colour bin (unless there is a mixed colour glass collection)

⇨ recycle your blue glass with your green glass

⇨ wash out bottles and jars, remove caps and corks before recycling (avoid wasting water: use your washing-up water)

⇨ light bulbs, Pyrex-type dishes, windowpanes etc. should not be put in glass banks.

Advice: Reuse jars for storage, most supermarkets have glass banks, recycle alongside your weekly shop!

Jewellery recycling

⇨ Recycle your jewellery...and help Marie Curie Cancer Care provide more nursing care at home to terminally ill people. All you need to do is put your jewellery in an envelope and send to FREEPOST, Central Recycling (no stamp needed)

⇨ alternatively you can take your jewellery into any Marie Curie Cancer Care shop or Laura Ashley store and drop it into the specialist box provided.

Advice: For more information please check out Marie Curie Cancer Care for further details.

Light bulbs

⇨ Light bulbs should not be put in glass banks

⇨ to find your nearest recycling point for Energy Saving light bulbs, enter your postcode into the bank locator search tool on the recycle-more website (www.recycle-more.co.uk/banklocator/banklocator.aspx) and tick the 'Gas Discharge Lamps' box. This will give you all the address details and an interactive map.

There are over 50 different types of plastics

Advice: If a low-energy bulb is smashed, the room needs to be vacated for at least 15 minutes. Do not use a vacuum cleaner to clear up the debris. Instead, use rubber gloves to place the broken bulb into a sealed plastic bag – and take it to the nearest appropriate recycling centre for disposal.

Medical waste

⇨ Dispose of medicines following either your doctor's or the manufacturer's instructions

⇨ care should be taken when disposing of needles and syringes

⇨ glass bottles and jars that have contained medicines can be recycled when they are empty.

Advice: If you find a syringe, use your common sense. If you can safely pick it up, then place it in a safe container and take it to the local police station.

Metals

⇨ Usually separated into: aluminium (drinks cans) – non-magnetic, and steel (food tins) – magnetic. Aerosols can be made from either

⇨ test by using a magnet.

⇨ wash and squash cans before recycling. Only recycle clean aluminium foil. Never pierce or crush aerosols even when empty

⇨ only put empty aerosols in recycling schemes i.e. when you cannot get any more out by pressing the button

⇨ crisp wrappers (metallised plastic film) cannot be recycled. Metallised plastic springs back when scrunched.

Advice: Contact Alupro for details of their Cash for Cans scheme and BAMA for any aerosol queries.

Paper and cardboard

⇨ Paper collection is usually separated into: newspapers, magazines, cardboard, and phone directories

⇨ unless specified, do not recycle catalogues, directories or envelopes which are gummed or glued together

⇨ juice and milk cartons cannot be recycled with ordinary paper as they are made up of several materials

⇨ some facilities provide mixed paper and card collection.

Advice: If you read newspapers, please recycle them after use. Alternatively, read news online. Set your printer to print double sided, buy recycled paper.

Plastic

⇨ There are over 50 different types of plastics

⇨ if separate bins are provided it will usually be for:
  ↳ HDPE - opaque bottles e.g. detergent bottles
  ↳ PVC - transparent bottles, an obvious seam running across the base e.g. mineral water bottles
  ↳ PET - transparent bottles, a hard moulded spot in the centre of the base e.g. fizzy drink bottles

⇨ some supermarkets have collection points for recycling carrier bags.

Advice: Reuse bags or use a long-life carrier bag. Buy in bulk to reduce packaging.

Rubber hot water bottles

⇨ Cut your hot water bottle into squares and use them as jam jar openers

⇨ or as non-slip mats for your plant pots

⇨ cut up your hot water bottle into

smaller strips and place them at the bottom of your plant pots for drainage

⇨ fill your hot water bottle with material such as reused foam chips and create a kneeler for your gardening.

Advice: Please do not burn your hot water bottles in your fire as they can produce harmful gases.

Stamps

⇨ There are many organisations that you can donate stamps to including: RNIB, Oxfam (also collect coins), Help the Aged and Guide Dogs for the Blind.

Advice: Don't forget to recycle your envelopes once you've donated all those stamps for charity!

Textiles

⇨ Old clothes, bedding, curtains, and blankets can be recycled on any high street at charity shops, but only donate clean usable items!

⇨ some charities also have recycling bins for textiles

⇨ if you deposit shoes, tie them together so they don't get separated!

Advice: Use any unrecyclable textiles as cloths around your home.

Timber/wood

⇨ The disposal of wood in landfill sites causes problems as it is often bulky and decomposes slowly

⇨ scrap wood is collected at civic amenity sites for recycling

⇨ contact your local council to find out whether they recycle Christmas trees.

Advice: Many retailers now stock products made out of recycled wood or renewable wood sources – look on the FSC website for further information.

Tools

⇨ Unwanted tools and equipment can be donated to Workaid who refurbish the items and send them to vocational training projects in developing countries. They accept any tools from handtools to typewriters to welding machines!

⇨ Tools For Self Reliance also refurbish quality second-hand tools to a first class standard

⇨ Tools With A Mission also collects tools for developing countries.

Advice: If your electric tools are beyond repair, check if your local tip recycles waste electrical equipment.

VHS and cassette tapes

⇨ You can post your old CDs, DVDs, CD-Rs, VHS and cassette tapes to The Laundry for recycling.

Advice: Reuse your VHS and cassette tapes at your nearest charity shop.

Water filters

⇨ All Brita Water Filters can be returned to the company to be recycled, please visit the Brita website for more details and the address: www.brita.net

Advice: If you use a different brand, contact the company direct to find out where to recycle the water filters.

⇨ The above information is reprinted with kind permission from Recycle More. Visit www.recycle-more.co.uk for more information.

© Recycle More

The machine that sorts out household rubbish

The days of having to sort out household rubbish before collection could be over after scientists created a machine to do it automatically.

Scientists have created the Autoclave system, which divides the waste for recycling on a huge scale and produces enough energy to power itself.

Its inventors, AeroThermal, say that at some point in the future, householders could even sell their rubbish because there is potentially a profit in it.

All steel and aluminium is cleaned during the process, plastics are reduced to recyclable pellets and glass is made reusable.

Food and organic refuse is turned into a bio-gas that can be converted into green electricity. Even the steam that is used in the process is recaptured afterwards and re-used so nothing is released into the atmosphere.

In two hours the technology, which acts like a giant steam-powered pressure cooker, can deal with 30 tonnes of municipal waste.

AeroThermal said the Autoclave system could be used to achieve the EU and government targets for dealing with our household waste.

Ian Toll, managing director of the firm based in Poole, Dorset, said: 'The system provides real answers to environmental problems.

'Disposal of rubbish poses headaches for authorities on a local and national scale, and we believe this system will reduce pollution – and the cost of waste management.

'It is proof that engineering and the application of science can go some way to help combat the major threats facing us today.

He added: 'Its effect will be felt by ordinary people because it means we could revert back to the old system of putting out rubbish – with everything in the same bag.

'It powers itself and there is enough green electricity left over to put back into the national grid – and it could ensure we reach recycling targets.

'The steel and aluminium is cleaned and all the labels are removed, and that increases its value.

'The plastics, including plastic bags, are separated and reduced so they can be recycled.

'And all the food and cellulose material is reduced to its basic form, and after it is put through an anaerobic digestion system it can be converted into electricity.

'There is no need for us to ship any of our waste to China when we have the technology to sort it out then recycle it.'

⇨ This article first appeared in the *Daily Mail*, 14 May 2008.

© 2008 Associated Newspapers Ltd

Recycle and reuse

What happens when we recycle

The facilities that you have available for recycling may be very different to those of your neighbours in the county next to where you live. It depends on where you live and on which Local Authority manages your waste and recycling collections as to what is collected, how it is collected and how often. You may not have access to a kerbside collection service and will then have to rely on bring banks and mini recycling centres. Because of these reasons, it is very difficult to offer specific advice about methods of recycling that will be relevant nationally – so the best thing to do is to visit your local council's website or telephone them directly to see what is available to you locally.

Having said that, however, below is a rough guide to some general rules about recycling and reducing waste. Information on what happens next to the items that you recycle have also been included because these are valuable resources. Nothing goes to waste if you recycle.

Paper

⇨ Separate paper from the other items in your recycling box (if you have one) to make it easier for the recycling collectors.

⇨ Most types of paper are accepted: newsprint, magazines, envelopes (including the plastic windows) and junk mail, without plastic wrappers.

⇨ With unwanted mail of course the better way is to stop it being delivered in the first place: contact the Mailing Preference Service (www.mpsonline.org.uk tel: 020 7291 3310) to have your name removed from mailing lists.

⇨ Coloured and brown paper were previously excluded because they can leave flecks in the finished product and reduce its brightness. However, processes are improving all the time and a wider range is becoming acceptable in some collections.

⇨ Every year we need a forest the size of Wales to provide all of the paper we use in Britain.

⇨ 33% of what we throw away is paper and cardboard.

What happens next?

Paper is sent to Belgium, or to Aylesford Newsprint in Kent, which is one of the largest recycling plants in Europe. All inks, glues, staples, plastic film etc. are washed out with soapy water, a process which is helped by the proportion of magazines in the mix. Magazines contain clays that help to lift inks during washing. Cleaned paper pulp is sent to a paper-making machine where it is injected between two wire meshes to form a damp sheet, before passing through hot drying cylinders. On Aylesford's production line, the paper is now moving at more than 60 mph as it rolls onto jumbo reels, each one about 30 tonnes in weight. This high quality newsprint supplies national and local newspapers throughout the UK and Europe. New papers could be coming back to you, in the newsagent's or through your door, within three to four weeks.

Yellow Pages

⇨ The dye in these directories makes them unsuitable for normal recycling. Also large numbers are discarded around the same time, as a new edition arrives, and so much material would taint batches of paper pulp.

⇨ Most collection services will take Yellow Pages through a box collection or from designated local collection points. Contact your local council for more information.

What happens next?

Yellow Pages are treated in a different way to other types of paper. Covers and glue are removed, pages are shredded and used in lots of imaginative ways: for animal bedding, Jiffy bags, cardboard and insulation for houses. An innovative scheme in Devon used shreddings beneath road surfaces to reduce noise. Near the Tewkesbury-based Highbed Paper Bedding company, some larger stables send used bedding for composting, so ensuring yet another 'life' and making maximum use of old Yellow Pages.

Cardboard

⇨ Cardboard packaging is everywhere. It can take up a lot of room in the average household rubbish. It is made of cellulose fibres, generally from wood pulp, which can be used again if recycled.

⇨ The UK produced an estimated 9.3 million tonnes of waste packaging in 2001. Of this 5.1 million tonnes came from households and the remaining 4.2 million tonnes from commercial and industrial sources.

⇨ Try to avoid buying items which contain large amounts of packaging. Some companies such as removal firms will supply cardboard boxes which they then take back for reuse.

⇨ Cardboard makes excellent compost. Scrunch it up and put it in your compost bin with kitchen and garden waste. It also makes excellent mulch for vegetable beds.

⇨ However, because of its light weight and low quality it holds little monetary value for recycling.

⇨ A variety of cardboard recycling collection schemes are in operation around the country. Some local authorities will collect from the doorstep using a box or bag collection service, others will accept a mixture of green waste and cardboard, and some local authorities take cardboard at local Household Waste Recycling Sites (council 'tips' or 'dumps') and others cannot collect them at all.

⇨ Cardboard containers need to be flattened as much as possible and empty. If possible, it is also helpful to remove any obvious fastenings, adhesive tape etc.

⇨ Contact your local council to find out what service is in operation in your area. Or find your nearest cardboard recycling bank using the online service www.recyclenow.com

What happens next?

Cardboard recycling involves soaking in water and agitating to release fibres, turning them back into pulp. Metal and ink contaminants are removed, additional finishing chemicals are added; the pulp is pressed into sheets and dried.

Although the fibres get shorter each time they are pulped, cardboard can be recycled four or five times before fibres degrade and disintegrate.

Second time around cardboard makes more boxes and packaging, but has an interesting range of other uses including stationery, animal bedding – and as a final resting place, coffins!

Glass

➪ 1,350,000 bottles and jars are recycled each month in Bristol alone.

➪ All bottles and jars are accepted, and it helps if they are rinsed, with caps and lids removed.

➪ If using bottle banks please sort bottles into the correct colours because if there is contamination with different colours of glass the quality of the glass is reduced. Blue bottles are classed as green.

➪ Only a few types of glass are not suitable, as they are manufactured differently, e.g. toughened (like Pyrex), window panes, and ornamental (such as vases).

➪ Glass is special because it can be recycled again and again. That means using less energy in furnaces, and fewer raw materials: EU law will soon demand that the UK recycle 70% of its glass.

➪ Buy refill packs or look for returnable bottles wherever possible.

➪ Reuse glass bottles and jars for storing odds and ends or donate to a local jam maker.

What happens next?

The glass is sorted by colour, washed and impurities are removed. It is crushed into cullet (small pieces) and melted, then moulded to make new bottles and jars. Glass can also be used as aggregate in road building: Glasphalt looks just like any other tarmac, but is 30% crushed glass, specially treated so it won't puncture tyres! Glass comes round again in more decorative ways too, for some walkways in Bristol city centre, for example, and graves were traditionally dressed with coloured glass chippings.

Food tins

➪ Food tins are made of steel, coated by a thin layer of tin.

➪ It is really important that food cans and tins are rinsed before collection: it is only a moment's work but very helpful. Not only are dirty tins unhygienic and unpleasant to deal with, contamination can disrupt the smelting process.

➪ You do not have to remove the labels from the tins as these are fired off during the extremely hot smelting process.

What happens next?

A magnet is used to separate the steel from aluminium cans. They are melted down in furnaces, with iron ore, and oxygen is added to remove impurities. The impure metal (slag) is separated and may be used in road-building. The pure metal is made into blocks (ingots), rolled into many shapes and sizes and water-cooled. It will be used for more tins, or car parts, fridges and other domestic appliances. On a grander scale, what once was a humble food tin might just become part of a bridge.

Aluminium cans

➪ Last year an estimated 5 billion aluminium cans were used in the UK.

➪ The energy it takes to make one new aluminium can is enough to make 20 recycled ones.

➪ This is the most valuable of recyclable materials.

➪ It takes 4 tonnes of bauxite to make 1 tonne of aluminium; mining and transport both use large amounts of energy.

➪ Aluminium is always in demand and it is very important to remove cans, and foil, from our waste bins.

➪ Aluminium cans should be cleaned before recycling, although the labels do not have to be removed as the hot smelting process will destroy these.

➪ Cash for cans schemes are run all over the UK where aluminium cans can be exchanged for cash donated

The energy it takes to make one new aluminium can is enough to make 20 recycled ones

to help raise funds for charities and other good causes.

➪ Find out more by contacting your Local Authority or visiting www.alupro.org

➪ You could use a can crusher to make storage easier.

What happens next?

Cans are sorted, baled and taken for crushing into large blocks, and sometimes shredded for reprocessing. Melting removes all inks and coatings before metal is made into blocks (ingots), which can be huge, 2 x 8 metres and 60cm thick, and weigh as much as 20 tonnes. Each one contains about 1.6 million drinks cans. Ingots are sent to mills where they are rolled into sheets from 0.006mm to 250mm gauge. This rolling adds strength to the pure aluminium which then travels far, to can makers all over Europe – and within just six weeks those new shiny drinks cans are back on the shelves.

Aluminium foil

➪ Bottle tops, take-away containers, as well as cooking and wrapping foil are all welcome.

➪ It is easy to mistake silver-coated plastic (such as crisp packets) for the real thing. The squash test works every time: aluminium foil will stay crushed in your hand, the plastic sort springs back.

➪ Foil should be washed and squashed together and kept separate in the recycling box.

What happens next?

Foil is recycled separately from cans because it is made from a slightly different alloy of metal. It is similar to the aluminium can process, without the de-coating or shredding. Ingots are much smaller, about a metre long, from which more foil is made, or a range of products such as light-weight car parts.

Clothes, shoes and textiles

⇨ 2 million pairs of shoes are discarded every week in the UK.

⇨ Different areas have their own range of materials which are accepted. Often the list is limited to wearable garments and shoes. Details can always be confirmed by contacting your own Local Authority.

⇨ Generally clothes should be reusable and clean.

⇨ Shoes should be tied in pairs.

⇨ It is important that these materials are kept dry to avoid mould which ruins them – one bag of damp clothes can contaminate a whole load.

⇨ Cloth and footwear should be carefully sealed in plastic bags, and never put out so much in advance of collection that they may get rained on!

What happens next?

Clothes and shoes are either sold to people here through charity shops, or are sent to developing countries where they are used again. The same applies to household linens, curtains etc. (where they are collected); lower quality textiles, not fit for wear, are taken in some districts and go for fillings or cleaning rags. Wool can be recovered and re-spun.

Spectacles

⇨ Unwanted glasses can be taken to either Dollond and Aitchison opticians or Help the Aged stores.

⇨ You can also send spectacles in good condition (they do not take broken frames or bifocals) to the charity Vision Aid Overseas, 12 The Bell Centre, Newton Road, Manor Royal, Crawley, West Sussex RH10 2FZ enclosing a compliments slip (so that they know who to thank).

⇨ Some Local Authority recycling collection schemes will take spectacles.

⇨ If using a recycling collection service it does not matter if the lenses or the frames are broken – donate them anyway.

What happens next?

They are sorted and cleaned and then passed onto a charity such as The World Sight Appeal or Vision Aid Oversees donate them to communities in developing countries.

VAO distribute them in developing countries, helping people who would not otherwise have access to any professional eye-care.

Car batteries

⇨ 80,000 tonnes of car batteries are thrown away every year.

⇨ Ask your Local Authority if you can recycle car batteries through your recycling collection service, or should you take them to your local Household Waste Recycling Centre (tip).

What happens next?

A huge press crushes the car batteries, breaking them down into valuable component parts which can then be carefully sorted:

Plastic is thoroughly washed, dried and ground up into granules which are used in many different products, including recycling collection boxes, furniture, paint trays, car parts, drainpipes and – fittingly – more car battery cases.

Lead is melted down to make not just more car batteries, but also guttering for roofs and shields for X-ray machines in hospitals.

Acid is treated and neutralised.

Distilled water is purified and used again.

Engine oil

⇨ 1 litre of oil can pollute a million litres of fresh drinking water.

⇨ Avoid spilling or burning – not only is this against the law but it can cause water and air pollution.

⇨ You should be able to recycle the oil through Household Waste Recycling Centres.

⇨ The Environment Agency have set up an oil care campaign to help oil users to dispose of oil responsibly. The Helpline provides advice and gives details of your nearest oil recycling bank. Alternatively this information can be found by calling 0800 663366 or go to www.oilbankline.org.uk

What happens next?

Containers of oil from household collections are decanted into large holding tanks. Oil is boiled and left to settle; any water is removed at this stage and the oil is filtered to remove metal particles. The process is repeated to produce a watery brown liquid

that is used in the furnaces at power stations, for heating tarmac and drying stone in quarries, as an alternative to conventional fuels.

Green waste

⇨ About a third of the average household refuse bin is made up of waste that could be composted.

⇨ Composting saves money – there's no need to fork out on commercial products from garden centres.

⇨ Composting cuts down on the need to buy peat-based products, and therefore saves our almost extinct peat bogs – these support rare plants and animals.

⇨ Home-made compost makes an excellent soil conditioner and a rich source of plant food.

⇨ It's easy AND it's free!

⇨ Ask your Local Authority if they supply subsidised home compost bins.

⇨ Garden waste can be taken to your local Household Waste Recycling Centre or you may receive a garden waste collection.

⇨ Find out details on how to recycle green waste by visiting www.wasteonline.org.uk and take a look at the composting fact sheet.

Green waste is vegetable matter, plant material, prunings, grass cuttings etc. from gardens. Green waste is not generally treated in the same way as anything that has been indoors in a kitchen environment, and which may have been near meat or fish, especially uncooked. When green waste is buried in landfill, there are potential problems with leachate (seeping liquid which pollutes the soil) and methane, a gas which is flammable and contributes to the greenhouse effect.

Composting is the best method of recycling biodegradable matter. Unlike the toxic cocktail of landfill, good composting conditions enable aerobic breakdown into nutrients and soil-conditioners, a valuable resource – and virtually free for gardeners. In some areas civic amenity sites compost green waste and offer it for sale to local people, or it may be used to enrich soil on farms.

To get started you can either:

⇨ Build a compost bin from old pallets or wood posts and wire mesh netting lined with old carpet or thick cardboard. Cover this with a wooden lid or old carpet to keep the rain out and heat in.

⇨ Or contact your Local Authority to find out if they sell subsidised compost bins.

Do compost:

⇨ Kitchen waste – such as fruit skins and vegetable peelings, tea bags, coffee grounds and crushed egg shells.

⇨ Garden waste – grass cuttings (but not too much at a time), hedge clippings, prunings, old plants and flowers.

⇨ Crumpled or shredded card and waste paper – including cardboard tubes and egg boxes. Try to avoid heavily coloured paper.

⇨ Wood ash – but not coal.

⇨ Human hair and animal fur.

⇨ Autumn leaves – in small amounts. Otherwise put them in bin liners where they rot down and are great for mulch.

⇨ Old pure wool jumpers and other natural fabrics.

⇨ Sawdust and bedding and manure from vegetarian pets such as rabbits.

Don't compost:

⇨ Cooked food, meat and fish.

⇨ Droppings from meat-eating animals.

⇨ Magazines and heavily inked cardboard.

⇨ Nappies.

⇨ Coal ash and soot.

⇨ Plants infected with persistent diseases such as clubroot, white rot and blight.

⇨ The roots of persistent weeds like bindweed or couch grass.

⇨ Synthetic fabrics.

⇨ Glass plastic and metal – these should be recycled separately.

What happens next?

Whatever you decide! Compost produced from your own compost bins can be used as a mulch to discourage weeds, dug into your soil around your plants or used in window boxes or pot plants.

Plastic

⇨ The world's annual consumption of plastic materials has increased from around 5 million tonnes in the 1950s to nearly 100 million tonnes today.

⇨ It is estimated that nearly 3 million tonnes of plastic waste is produced a year in the UK.

⇨ Wherever possible buy refillable plastic containers and try to avoid unnecessary packaging

Plastic is a problem, and most people realise why. It is not going to go away: because natural processes will never be able to break it down. Its manufacture uses petrochemicals from oil supplies which cannot be replaced, and involves high-temperature furnaces and long-distance travel. Plastic is also very light, often filled with air, and can take up a huge amount of room. Most discarded plastic is buried in landfill. But it is valuable and should have more than one life – above ground!

There are many different types which must be separated before processing and the 'bottle' type is most suitable for recycling. So the kind of container used for milk, fizzy drinks, shampoos, detergents, cleaning fluids etc., is collected. At present it is not possible to accept plastic film or carriers, tubs and pots or the sort of punnet in which fruit and meat is sold.

Two main types of plastic are recycled: basically clear and opaque. These are chopped into flakes, formed into pellets, then melted down for manufacture into various new products – although the material will not be used to contain food or drink again. Instead hard surfaces for furniture are made or flexible drainage pipes; most inspiring of all is the high quality fleece which can be produced for outdoor clothing.

Some areas are lucky enough to have a kerbside collection for plastic bottles, in others they have to be taken back to the supermarket (some Tescos' and Sainsbury's provide huge containers

to make it easy); Civic Amenity sites also offer facilities for plastic recycling. Bottles – with tops removed – need to be rinsed, and flattened to save space. (It can be fun, squashing bottles flat, and children are usually willing to help!)

It's worth the effort, and more people are understanding why. Any contribution, however small, will mean a little less plastic buried for ever!

Nappies

Throwaway nappies are costing the earth – literally.

For tomorrow's world and today's children: it's time to rethink.

Set up as a waste minimisation initiative in 2001, The Real Nappy Project encourages parents, nurseries, clinics and hospitals to use washable nappies and reduce the volume of disposables going into the waste stream. It is run by the Recycling Consortium; an awareness-raising not-for-profit organisation.

For further information on The Real Nappy Project, please visit www.recyclingconsortium.org.uk/community/nappies.htm

Computers

Every year over 1 million computers end up in our landfill sites. At the moment less than 20% of old computers are recycled! There are a number of national companies which take large amounts of redundant PCs from businesses for reuse, as well as local community projects which take PCs for refurbishment and then pass these on to charities, schools, low income households and developing countries overseas.

This material was produced in 2005 for the Bristol area and should be read with that in mind. In 2006 The Recycling Consortium was part of a merger which resulted in the formation of Resource Futures. Resource Futures continue to offer waste education services in Bristol, South Gloucestershire, North Somerset and Devon and can provide advice to local authorities wishing to develop curriculum based waste education services in schools.

⇨ The above information is reprinted with kind permission from Resource Futures. Visit www.resourcefutures.co.uk for more information.

© *Resource Futures*

Steps to successful home composting

A step-by-step guide

Composting is Nature's way of recycling and helps to reduce the amount of waste we put out for the bin men. By composting kitchen and garden waste you can easily improve the quality of your soil and be well on your way to a more beautiful garden. The following easy guide to home composting will provide you with all the information needed to get the best out of your bin. Now let's get composting!

Step one – placing your bin

It's best to site your bin on a level, well-drained spot. This allows excess water to drain out and makes it easier for helpful creatures such as worms to get in and get working on breaking down the contents. Placing your bin in a partially sunny spot can help speed up the composting process.

Step two – put these in

Like any recipe, your compost relies on the right ingredients to make it work. Good things you can compost include vegetable peelings, fruit waste, teabags, plant prunings and grass cuttings. These are considered

'Greens'. Greens are quick to rot and they provide important nitrogen and moisture. Other things you can compost include cardboard egg boxes, scrunched-up paper and fallen leaves. These are considered 'Browns' and are slower to rot. They provide fibre and carbon and also allow important air pockets to form in the mixture. Crushed eggshells can be included to add useful minerals.

Step three – keep these out

Certain things should never be placed in your bin. No cooked vegetables, no meat, no dairy products, no diseased plants, and definitely no dog poo or cat litter, or baby's nappies. Putting these in your bin can encourage unwanted pests and can also create odour. Also avoid composting perennial weeds (such as dandelions and thistle) or

weeds with seed heads. Remember that plastics, glass and metals are not suitable for composting and should be recycled separately.

Composting is Nature's way of recycling

Step four – making good compost

The key to good compost lies in getting the mix right. You need to keep your Greens and Browns properly balanced. If your compost is too wet, add more Browns. If it's too dry, add some Greens. Making sure there is enough air in the mixture is also important. Adding scrunched-up bits of cardboard is a simple way to create air pockets that will help keep your compost healthy. Air can also be added by mixing the contents. After approximately 6-9 months your finished compost will be ready.

Step five – using your compost

Finished compost is a dark brown, almost black soil-like layer that you'll find at the bottom of your bin. It has a spongy texture and is rich in nutrients. Some bins have a small hatch at the bottom that you can remove to get at the finished product, but sometimes it's even easier to lift the bin or to tip it over to get at your compost. Spreading the finished compost into your flowerbeds greatly improves soil quality by helping it retain moisture and suppressing weeds. Composting is the easiest way to make your garden grow more beautiful.

⇨ The above information is reprinted with kind permission from Recycle Now. Visit www.recyclenow.com for more information.

© Recycle Now

How green are we?

New figures show Britons back recycling

By Karen McVeigh

The first signs of a green revolution are emerging around the country as Britons treble their recycling and increase their use of public transport. But this enthusiasm is not reflected in attitudes towards other environmental concerns, with car ownership and use on the rise and air travel increasing 'substantially' over the last four years.

Recycling increased by 27% across all regions between 2002 and 2006

The Office for National Statistics said yesterday that research shows the public sending out mixed messages on its green credentials, although attitudes to the environment are changing.

Paul Vickers, deputy head of regional statistics at the ONS, said there had been a definite change in behaviour. 'The figures show that we are all recycling more. There are many reasons for this; one of them is local authorities and how aggressive they are in introducing policies to meet government targets on waste. But it also reflects a change in people's attitudes.'

Recycling increased by 27% across all regions between 2002 and 2006, with households in the east of England recording the highest figure of 34%, almost double the rate of four years previously. London and the north-east recycled the lowest proportion.

The amount of household waste produced over the last four years has remained broadly the same. London was found to produce the least waste per household, at 21kg a week between 2005 and 2006, due to the number of people living alone. By contrast, households in Northern Ireland generated the most waste at 26kg, due to larger than average household sizes. The government

has set a target for 40% of waste to be recycled by 2010. Separate figures for 2007-08, released yesterday by the environment department, show that the recycling trend has continued this year, to 33%, with a fall in the amount of waste going to landfill.

Vickers said that the attitude shift towards recycling contrasted strongly with the rise in air and car travel seen over the same period. He said: 'Car travel is on the increase, as is air travel. What we have seen is a growth in the use of regional airports, particularly Bristol, Liverpool and Southampton, where passenger numbers have doubled. Cheap airlines seem to be driving this trend. The number of cars has gone up, and car use also, but the number of new registrations since 2001 has fallen. This could be because people are hanging on to their cars longer.'

During the period 2003-06, the average distance travelled by drivers increased by over 3% to nearly 5,900 miles a year. People in the south-west travelled the furthest, an average of 7,100 miles in 2005-06. The number of licensed cars in Britain has grown steadily in the 10 years to 2006, to 28m. The north-east and the

East Midlands showed the largest increases, with 30%, compared with just 9% in London.

But while car use has increased overall, it has decreased in London, by 2%, and in the East Midlands, by 7%. Both these drops in use were accompanied by an increase in the use of public transport, by 23% in London and 29% in the east Midlands.

Londoners were found to be some of the greenest citizens in the UK, walking the furthest, an average of 230 miles, and cycling the furthest, an average of 50 miles, in 2005-06. 'We have looked at London as a whole during the periods 1991-2001 and 2005-06 and we have found that the distance travelled by car has dropped by 15%,' said Vickers. 'Over the same time period, the distance travelled by public transport has increased by 30% and the distance cycled has increased by two-thirds.'

The results prompted the ONS to question whether the introduction of the congestion charge was causing the shift in London, but it was unable to reach a conclusion. 'Because our figures were for the whole of London and not just the congestion charge zone we couldn't draw conclusions.'

Car travel far outstripped journeys made by other means of transport, but rail travel increased between 2003 and

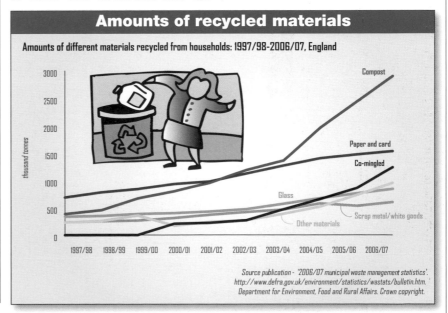

Amounts of recycled materials

Amounts of different materials recycled from households: 1997/98-2006/07, England

Compost
Paper and card
Co-mingled
Glass
Other materials
Scrap metal/white goods

Source publication - '2006/07 municipal waste management statistics'.
http://www.defra.gov.uk/environment/statistics/wastats/bulletin.htm.
Department for Environment, Food and Rural Affairs. Crown copyright.

2006. The Yorkshire and the Humber region showed a 93% growth in rail travel, followed by the East Midlands, with a 73% rise. Rail travel in the West Midlands declined by 37%.

The government has set a target for 40% of waste to be recycled by 2010

All the UK's major airports showed a rise in the number of flights between 2001 and 2006. Stansted dealt with 23.7 million passengers in 2006, an increase of 73% since 2001, as travellers took advantage of budget airlines. Regional airports such as Liverpool, Bristol and Southampton all doubled their passenger numbers within the five-year period.

Paul Bettison, chairman of the Local Government Association's environment board, said: 'It is very encouraging to see people doing their bit for the environment by recycling more and reducing this country's reliance on landfill.'

But he said that while the figures were a move in the right direction, there was a need to do more. 'Councils and council tax payers are still facing fines of up to £3bn if we do not dramatically reduce the amount of waste thrown into landfill,' he said.

Greenpeace transport campaigner Anna Jones said: 'It's no surprise that the number of flights taking off from UK airports is going up when we're led by a government that has done more than any other to promote a binge-flying culture. Ministers are backing new airports and new runways when they should be making trains cheaper and more accessible.'

Friends of the Earth's head of campaigns, Mike Childs, said ministers had to do much more to help people live 'less polluting lives', including tougher energy efficiency standards for products and cars, greater investment in public transport, and taxes to make it cheaper and easier for people to go green.
9 May 2008

© Guardian Newspapers Limited 2008

Hazardous landfill waste falls

Environment Agency celebrates 60 per cent fall in the amount of hazardous waste going to landfill but new enforcement policies are on the horizon

New waste regulations such as the Landfill and Waste Electrical and Electronic Equipment (WEEE) directives are having a positive impact, according to new figures from the Environment Agency.

The data, released last week, revealed that 60 per cent less hazardous waste was sent to landfill last year than in 2004. Levels of recycling also increased over the same period, with 50 per cent more potentially dangerous waste being recycled or re-used.

Martin Brocklehurst, head of external programmes at the Environment Agency, said that new regulations governing hazardous waste had been instrumental in the improvements. 'As new legislation like the Landfill directive and Waste Electronic and Electrical Equipment (WEEE) regulations kick in, we're starting to see a shift in how we deal with hazardous waste,' he said. 'Business and industry are adapting to the changes.'

The research said it was less clear as to whether or not manufacturers were successfully removing hazardous substances from their products, noting that while the amount of waste classified as hazardous rose 12

---

60 per cent less hazardous waste was sent to landfill last year than in 2004. Levels of recycling also increased over the same period

per cent between 2004 and 2006, this was partly because of changes in the rules on hazardous waste. It also found that the key business sectors such as the oil and solvents industry had produced less hazardous waste over the period.

The findings came as the Environment Agency hinted it would step up its policing of producers and handlers of hazardous waste. The Agency said it was currently revising its guidance and enforcement priorities, adding that its new guidance 'puts the onus on producers of hazardous waste to ensure their outputs are properly classified and treated'.

The Agency said it would also focus its enforcement efforts on those firms guilty of mis-describing hazardous waste. It added that by June next year it will expect all mixing of hazardous waste during treatment to have ceased and for all outputs from hazardous waste treatment facilities to be classified and coded in line with the Classification and Coding Guidance.
4 December 2007

⇨ The above information is reprinted with kind permission from BusinessGreen. Visit www.businessgreen.com for more information. The original article can be found at: www.businessgreen.com/2204975

© Incisive Media 2008

Landfill sites have a green future

Information from the Department for Communities and Local Government

Restoring landfill sites by turning them into green space, such as woodland, parkland or farmland is now possible, new research published today shows.

Many local people find landfill sites detrimental to their local area and a common solution is for councils to close them over with a compacted clay cap to seal up the waste. There are about 2,500 closed and operational landfill sites in England and Wales.

The results of a 10-year research project into the establishment of woodland on landfill show that it is possible to restore these areas safely by planting certain trees as long as strict safeguards are adhered to.

The Government is committed to reducing the UK's reliance on landfill to reduce their environmental impact. The Landfill Directive has promoted more sustainable solutions and brought in important regulations that include abolishing the disposal of liquid, clinical and other hazardous waste.

Establishing trees and woodland on landfill has previously presented real challenges for landfill operators and local authorities, and until recently government guidelines actively discouraged it because of fears that the tree roots might not grow deep enough and if they did they might pierce the 'cap' letting out landfill gases.

In 1993 the Government acknowledged that further evidence and reassurances were needed to determine whether this could be done safely. The Forestry Commission were asked to establish and monitor a number of experimental sites, which were specially engineered to control pollution with the dense compacted landfill cap with a thick layer of soil for the tree roots.

It is possible to restore [landfill] areas safely by planting certain trees

---

Today's Forest Research report, funded by Communities and Local Government, has found that good tree growth on these landfill sites has been achieved and that the establishment of vegetation is a vital part of their restoration. Woodland planting can now be recommended as long as specific site safeguards including that the underlying mineral cap is constructed to standards required by government guidance. Poplar, alder, cherry, whitebeam, oak, ash and Corsican pine have been identified as well suited to the landfill environment.

Planning Minister Iain Wright said:
'Many people find landfill sites a local eyesore and the government is committed to reducing landfill use.

'This new research shows that with the proper safeguards in place we can reduce the impact of old sites by planting them and environmentally reviving them as attractive woodland or parkland.

'Restoring landfill sites in this way can provide local communities with more attractive green spaces, help tackle climate change, regenerate important brownfield land and provide new places for wildlife to live.'

Professor Andy Moffat from Forest Research, the Forestry Commission's scientific and research agency, said:

'Waste management and dealing with waste disposal sites such as landfills once they have reached their capacity, are significant environmental challenges, and restoring them to woodland is an attractive option in many cases.

'There is still further research to do particularly on long-term performance of trees on landfill sites and the specifications of soil caps, but as a result of this research we believe that with careful planning and management many landfill areas can be successfully restored as woodland.'
30 May 2008

⇨ The above information is reprinted with kind permission from the Department for Communities and Local Government. Visit www.communities.gov.uk for more.

Recycling is not enough – we need to consume less

Information from the Economic and Social Research Council

Recycling rates have risen, and the UK is on schedule to meet EU targets, but the key to dealing with our escalating waste problem lies in changing our buying habits and our attitudes to consumption, according to the authors of a new Economic and Social Research Council (ESRC) publication.

Consumption: reducing, reusing and recycling, which accompanied a seminar in Belfast organised jointly with the Office of the First Minister and Deputy First Minister, Northern Ireland, says that the benefits of recycling risk being undermined by the sheer quantity of waste being generated. If household waste output continues to rise by three per cent a year, the cost to the economy will be £3.2 billion and the amount of harmful methane emissions will double by 2020.

The report highlights the many ways that social science can contribute to waste policy development, either by devising initiatives, by providing tools to evaluate their relative effectiveness or by helping understand why they did or did not work.

Professor Ken Peattie, Director of the ESRC Centre for Business Relationships, Accountability, Sustainability and Society (BRASS), Cardiff University, describes three projects which are linked to different aspects of waste reduction at the production stage and in consumption. He says the key tool in the development and implementation of consumption reduction policies is 'social marketing', which involves using commercial marketing techniques to influence their behaviour for the benefit of society as a whole.

Ken Peattie explains that social marketing can be successful because it focuses on the target audience's point of view, taking account of any emotional or physical barriers that may prevent people from changing their behaviour. 'Guilt messages are ineffective. A focus on the benefits of a greener lifestyle has been shown to be a better way to encourage people to reduce their consumption,' the report says.

Ben Shaw, Senior Research Fellow, Environment Group, Policy Studies Institute, describes international efforts to become more resource efficient by significantly reducing waste or achieving higher rates of recycling or reuse. He says that despite recent improvements the UK is still a long way behind the best performing countries and regions where taxation and household waste charges have been used to reduce landfill.

However, even the toughest penalties have not been enough to prevent a significant accumulation of waste. Ben Shaw says that waste reduction needs to be tackled higher up the chain of production and consumption. 'Waste reduction must be a goal of UK environmental policy, and not tackled through waste policy alone,' he says.

The report also gives examples of zero-waste initiatives which have been tried – from the high-tech, large-scale waste management systems of consumerist San Francisco, to the locally based, small-scale initiatives in the Philippines.

Ben Shaw says that although there are some inherent problems with 'zero waste' as a concept and as a policy objective, there are lessons to be learnt by critically considering the achievements of existing practice, wherever in the world that may be found. For example:

➪ We should set a per capita residual waste target to drive both recycling and prevention, backed up by variable charging of householders.

➪ We should be among the first countries to tackle consumption by making innovative and transformative producer responsibility agreements.

➪ We could be much more ambitious in our recycling targets. We should try harder on construction and demolition waste.

➪ We should develop more 'closed loop' systems for organic wastes, for instance by returning composted food waste to the land as fertiliser, rather than losing this valuable resource.

15 June 2007

➪ The above information is reprinted with kind permission from the Economic and Social Research Council. Visit www.esrcsocietytoday.ac.uk for more information.

© *Economic and Social Research Council*

Why recycling isn't really saving the planet

Richard Wellings writes on environmental issues for the *Yorkshire Post*

The government constantly tells us that recycling is a wonderful idea. TV advertisements bombard us with this message. Children are taught to recycle at school. But are the environmental benefits of recycling really worth the economic cost?

That cost is growing rapidly. The landfill tax, introduced to encourage recycling, will increase by an inflation-busting 14 per cent in April. This levy will cost UK businesses and householders £900m in 2007.

Then there are the costs of collecting and reprocessing the recyclable materials – an estimated £400m per year, paid for by council tax payers.

Numerous EU waste directives are also being forced on different industries. The Waste Electronic and Electrical Equipment Directive alone cost British business £200m in 2006.

Despite all this expenditure, running into billions, it is difficult to identify any significant benefits from recycling – except where it occurs commercially without government intervention, as in the traditional scrap metal business.

The amount of municipal waste being sent to landfill has declined, but then again modern landfill sites have few negative environmental impacts. They are lined so that pollution cannot seep into water courses. Unpleasant odours and dust are carefully controlled.

Often landfill has a positive role in rehabilitating landscapes scarred by quarrying, such as the Yorkshire Wolds.

It is a myth that there is no more room for landfill. Currently, landfill sites take up a little over one-tenth of one per cent of the UK. Since rubbish rots and can be piled up, there is no practical limit on the amount of waste that can be stored. Any shortage of landfill capacity has been artificially created by the planning system and a government desperate to provide a rationale for costly recycling policies.

Another argument is that recycling saves energy and therefore reduces carbon emissions. Yet the amount saved is trivial, and there are far more efficient ways of reducing consumption – for example, by ending subsidies to almost empty trains and buses.

Even if every steel can in the UK was recycled the amount saved would amount to less than one-thousandth of the country's annual energy usage.

Many people also believe that recycling paper saves trees. In fact, the opposite is true. Recycling reduces the demand for wood pulp and reduces the value of trees. Fewer are therefore planted. This is a shame because young forests are particularly good at absorbing greenhouse gases.

Recycling may also be having a damaging effect on many Third World countries. Some, such as Jamaica, are heavily reliant on exports of raw materials such as bauxite (to make aluminium) and iron ore (to make steel). By subsidising recycling, the Government is indulging in a form of protectionism, effectively reducing the demand for imported raw materials while artificially supporting the domestic waste-processing industry.

Poverty is the real enemy of a clean environment. In Africa, tens of thousands die every year from respiratory diseases caused by burning wood and dung in open fires for cooking.

Recycling both reduces the income of poor countries that rely on raw material exports and reduces economic growth in industrialised countries burdened with additional taxation and red tape.

Given that the environmental case is weak and that the economic case is non-existent, why does the Government continue to promote recycling?

One answer is that it doesn't have much choice. EU legislation means that unless the proportion of household waste recycled is increased dramatically, local councils will face huge fines of up to £150 per tonne. This is particularly worrying for boroughs such as Bradford and Scarborough, where recycling rates are well below the national average.

Within local authorities there are also those with much to gain from the growth in recycling. A growing army of bureaucrats has been employed to supervise the implementation of EU targets by microchipping wheelie bins and rummaging through other people's rubbish – a Soviet-style nightmare that few could have envisaged even five years ago.

Many waste-processing companies have also gained from recent legislation, and these comprise an influential lobby in favour of further protection and subsidy for their industry.

The main losers are ordinary householders facing increases in council tax, higher prices in the shops and being forced to perform the pointless chore of separating out their tin cans and newspapers.

With the introduction of hefty fines for those who refuse to recycle, it remains to be seen how much state control will be necessary to enforce these costly and misguided policies.

Richard Wellings is Deputy Editorial Director at the Institute of Economic Affairs

⇨ The above information is reprinted with kind permission from the Institute of Economic Affairs and the *Yorkshire Post*, where it was first published on 4 April 2007. Visit www.iea.org.uk for more information.

Government ready to act on plastic bags

Information from Number 10

Gordon Brown has said that the Government is ready to take the 'necessary steps' to bring about a reduction in Britain's plastic bag consumption in an effort to improve the environment and cut pollution.

Writing in the *Daily Mail* newspaper today, Mr Brown said it was time for the Government, individuals and supermarkets to 'accept our own responsibility for ending the environmental damage we are causing'.

He said:

'I want to make it clear that if Government compulsion is needed to make the change, we will take the necessary steps. We do not take such steps lightly – but the damage that single-use plastic bags inflict on the environment is such that strong action must be taken.'

The Prime Minister praised the example of retailers like Ikea, which removed single-use plastic bags from its stores last July, and Marks & Spencer, which yesterday announced it will charge 5p for each one issued.

In November, Mr Brown held a forum with retailers to discuss how the 13 billion plastic bags given out to UK shoppers every year can be eliminated. In today's article the PM went a step further and warned that legislation could be enacted to bring about change.

The Government is ready to take the 'necessary steps' to bring about a reduction in Britain's plastic bag consumption

Mr Brown added that despite taking steps to reduce his carbon footprint, such as installing solar panels at his Scottish home and using low-energy electrical appliances, his family found themselves with 'a bin full of plastic bags' from supermarket deliveries. Such disposable bags represent 'one of the most visible and most easily reducible forms of waste', he said.

He continued:

'My approach is this: if we want others to change, we must make it easier for people to make the right

choices. That applies to individual things each of us do, and also what our firms and shops do. So the Government is ready to do what it can to encourage a change in the way we use these bags.'

Mr Brown said he would look at initiatives from around the world before deciding how to push ahead with the next steps on eliminating disposable bags.
29 February 2008

⇨ The above information is reprinted with kind permission from Number 10. Visit www.number10.gov.uk for more information.

© Crown copyright

Consumers oppose nanny state on plastic bags

Information from Ipsos MORI

In the wake of possible impending legislation which will force retailers to charge for plastic bags, new research from Ipsos MORI indicates that the government may face consumer opposition to their plans – particularly from less affluent shoppers who resent being dictated to on this issue.

Although 57% of the population agrees it is right for retailers to start charging for plastic shopping bags, 52% of people feel they should be able to make up their own minds about this issue and not have it forced on them by retailers wanting to charge. 35% feel it is wrong for the government to ban retailers from giving them away free.

Significantly it is Londoners who are most anti the prospect of having to pay for plastic shopping bags. Also, of the 22% who agree that they would prefer to pay for a bag than bring their own, a higher proportion than average are young shoppers aged 15-24 and affluent shoppers.

Comments Gill Aitchison, President, Global Shopper & Retail Research, Ipsos MORI: 'Our findings indicate that although people pretty much accept the concept of being charged for carrier bags, they don't appreciate being forced to accept this either by retailers or the government. Unsurprisingly, the move is least popular with less affluent consumers as currently they often use carrier bags as substitute bin bags – while for Londoners the idea of having to take a shopping bag with you when you go shopping can present a practical problem.'

52% of people feel they should be able to make up their own minds about this issue and not have it forced on them by retailers wanting to charge

'It is interesting to note the attitudes of the young and wealthy on this topic. While conventional thinking might suggest this group is more concerned with protecting the environment, they are not apparently willing to sacrifice the convenience of plastic bags and would rather pay than bring their own when they go shopping.'

When it comes to the proceeds of funds raised through charging for carrier bags, views of the public seem very clear. 74% of the population feel it is right for money raised to go to charity – this rises to 84% for 15- to 24-year-olds and is shared even by those who disagree with the principle of charging for bags (76%).

Concludes Gill Aitchison: 'While many consumers seem wedded to the convenience of free plastic bags, they do seem to have a strong social conscience when it comes to allocating the proceeds of any charges made. This indicates that widespread adoption of schemes such as those recently announced by M&S and Budgens may be a useful dimension to retailers' CSR policies in the future.

In addition, given the relatively high number of respondents who said that they would rather bring their own bag than pay for one (59%), there are signs that the Anya Hindmarch re-usable shopping bag trend is set to intensify.'

Technical note

This research was conducted by Ipsos MORI between 7th and 13th March 2008 via CAPIBUS, the face-to-face omnibus. Data collection took place in home with the respondents and 986 interviews were conducted amongst a nationally representative sample of GB adults aged 15+.
13 March 2008

⇨ The above information is reprinted with kind permission from Ipsos MORI. Visit www.ipsos-mori.com for more information.

Waste exports

Better protection for developing countries

Developing countries will be better protected against receiving unwanted waste from wealthier nations under revised international rules.

The 'Green List' regulation – which covers the export of non-hazardous recyclable materials from the EU – has been updated to formally record the wishes of countries outside the Organisation for Economic Cooperation and Development (OECD) that have expressed an opinion about the recyclable materials they would like to receive. Where a country has not expressed an opinion, agreement must be given on a case-by-case basis.

The aim is to protect these countries from receiving materials they do not want, and the change will also help prevent the export of recyclables to countries which believe they could not process them in an environmentally sound way.

Complementary changes to the Transfrontier Shipment of Waste Regulations will come into effect on 5 February this year, and will make it an offence for anyone to export material against the specified wishes of non-OECD countries.

Environment Minister Joan Ruddock said:

'Many developing countries want our recyclables because of the value to their growing manufacturing sectors. There is a double environmental win from this trade – it makes more sustainable use of the world's resources, cutting the consumption of virgin raw materials, while boosting recycling levels in the UK and reducing our reliance on landfill.

'But it is essential that this important legitimate trade is carried out in a mutually respectful and beneficial way. It is completely unacceptable to use it as a cover for dumping unwanted materials on countries that have no use for them, or cannot process them efficiently. This revised regulation will help prevent that happening.

'Defra has worked closely with other UK authorities, the European Commission, and other Member States, to ensure that the new Green List regulation is as complete and accurate as possible.

'I am very pleased that it now reflects the current wishes of countries to which we have a clear duty of responsibility.'
11 January 2008

⇨ The above information is reprinted with kind permission from Defra. Visit www.defra.gov.uk for more information.

Time to waste – tackling the landfill challenge

New report confirms pitfalls of charging for rubbish – but urges bravery on energy from waste

The report:

⇨ Highlights drastic figures that show most rubbish in the UK is still not recycled;

⇨ Recommends a major increase of energy-from-waste incineration as best way to stave off landfill crisis;

⇨ Believes the looming landfill tax burden must force a rethink to protect the council tax payer;

⇨ Suggests Government should use community-based incentives to encourage recycling, rather than household rewards and charging.

Time to Waste confirms that Government plans to scrap charging for rubbish would be a wise move, but warns that other tough decisions have to be made if the UK is to reduce the amount of waste it throws out and avoid hefty landfill burdens on council tax payers.

Citing research that currently only 20% of all household rubbish thrown out in Britain is recycled – one of the worst rates in Europe – the New Local Government Network (NLGN) has called for councils to introduce financial incentives for people to reduce the amount of waste they produce, instead of threatening them with fines. Under the alternative plan, neighbourhoods would be rewarded for reducing their waste output and increasing the amount they recycle with grants that they can spend on their local community. The money could be spent on items such as better street lighting or children's equipment for a local park.

The report concludes that 'introducing penalising charges for waste will not assist in the amount of rubbish society throws away' and highlights concerns that 'charging for rubbish would be unpopular, difficult to administer and could increase fly-tipping'.

However, the report argues for bravery on new alternatives to deal with the waste challenge, in particular the need to radically increase carbon-friendly energy-from-waste incineration. The report acknowledges that energy-from-waste incineration plants can provoke instinctive unpopularity in some local communities and therefore recommends returning some of the financial benefits from energy sold back to the grid to neighbouring residents, perhaps offering a £50 discount on the energy bills of households within those areas that agree to host the new plants. The report suggests that to meet the EU target to reduce the amount of land-filled waste, 10 large-scale or 200 smaller-scale plants may have to be built.

Britain currently produces around 330 million tonnes of waste each year and disposes of the majority of it into landfill. Figures show that Britain sends 7 million tonnes more rubbish than any other European country. It is estimated that the country will run out of landfill space in around nine years, with London and the South East due to run out in four years. The NLGN report therefore argues that the Government should shift its emphasis onto creating energy-from-waste through a new generation of incineration plants.

It also argues that councils could do more to inform the public about the need to reduce the amount of waste they throw out, such as publishing how much waste it disposes of each year and how much is recycled, alongside indicating waste charges separately on council tax bills to highlight how much the council spends on it.

The report is supported by Serco and United Utilities.

NLGN Director Chris Leslie said:

'With landfill tax increasing year on year and some authorities spending millions of pounds dealing with the problem of rubbish disposal, it is clear that the Government has to adopt a new approach to this challenge. None of the options we looked at are simple but switching away from buying rubbish to creating energy from waste is the greenest, most efficient solution.'

'We also want to see households given a positive incentive to reduce the amount of rubbish they throw away, rather than being persecuted by individual fines. By offering local communities financial incentives, residents would be able to benefit their locality as well as the wider environment.'

Mike Boult, Managing Director of Serco, said:

'There will come a point where the cost of waste disposal will become so high that it will make energy from waste profitable. There is a perception in the private sector that the financial benefits will, in the long term, enable you to get cheaper electricity...and if the council is saving on waste disposal, it could translate into council tax reductions.'

Time to Waste: Tackling the landfill challenge by Giorgia Iacopini, published 8 May 2008.

⇨ The above information is reprinted with kind permission from the New Local Government Network. Visit www.nlgn.org.uk for more information.

Renewable energy from rubbish is possible

By Andrew Hamilton

The European Commission recently set the ambitious target of producing 20 per cent of Europe's energy from renewable sources by 2020.

Europe as a whole has achieved roughly a third of this already. The UK, on the other hand, is seriously lagging. At present it gets just 2 per cent of its electricity from renewables and the Government has said it is unlikely to reach more than 16 per cent at best by 2020.

A separate, and yet no less pressing concern is that of the mounting piles of rubbish in landfill sites scattered across the country. Despite the concerted push towards recycling, more than 70 per cent of the UK's waste ends up in landfill sites.

This situation will soon become untenable when the European Landfill Directive bites harder, driving the cost of dumping rubbish beyond the budget of many local authorities.

That rubbish could be put to use in waste-to-energy schemes, simultaneously boosting the UK's energy output from renewable sources, lowering emissions from energy production, and drastically cutting the amount of waste we send to landfill.

Research from the Institute of Civil Engineers suggests that the UK could generate 17 per cent of its total electricity needs from waste, making it a credible, lower cost low-carbon alternative to nuclear energy.

A common misconception is that incineration is the only method of extracting energy from waste. This process has been widely criticised as inefficient and polluting. It involves transporting rubbish across the country to burn in vast incineration plants, and is incompatible with recycling schemes.

It also produces huge mounds of ash – up to 25pc of the volume of rubbish burnt – which then has to be dumped in landfill sites.

In fact, there are a number of alternative technologies for the conversion of waste without incineration, which work in conjunction with recycling schemes. Plants can be built to a size suitable for local communities, avoiding the cost and emissions associated with transporting waste around huge catchment areas.

A recent study by environmental consultancy Eunomia for the Greater London Authority examined the carbon footprint of a range of these technologies, showing that electricity and heat can be produced from waste with minimal carbon emissions.

Waste generates a steady flow of power, making it easier to feed into the grid than the intermittent bursts of wind energy. And its output is far greater. The plants are significantly less intrusive than wind farms – Advanced Plasma Power's facilities are no bigger than a distribution warehouse – so should not suffer the same lengthy delays in the planning process.

Local authorities have – with some prodding from Government – traditionally favoured large private finance initiative (PFI) contracts to manage their waste, which can lock them in for as long as 25 years. Small waste-to-energy companies offer shorter contracts to design, build, finance and operate plants.

They require no down payment from local authorities and the projects pay for themselves through a mixture of gate fees and the sale of electricity. This is of particular interest to the major energy providers, who must boost their renewable energy output from current levels of 1.3 per cent to 15 per cent to meet the European targets.

With commercial waste predicted to increase by 50 per cent by 2020, the ability to build small plants in industrial areas could transform the energy landscape with local waste being used to produce local electricity.

It is this decentralised approach that the UK should pursue to meet its energy and waste management goals simultaneously. It could even be the Government's first genuine piece of joined-up thinking.

Andrew Hamilton is chief executive of Advanced Plasma Power. The company uses a plasma gasification process to turn waste into energy using high electrical energy and high temperature (1,400°C – 2,000°C). The waste is broken down into gases and solid waste. After cleaning the gas can be used for power. The solid waste becomes an aggregate which is harder than granite and is used in the construction industry. www.advancedplasmapower.com
20 February 2008

KEY FACTS

⇨ The total amount of UK household waste in 2005/06 was 28.7 million tonnes, down 3 per cent from 29.6 million tonnes in 2004/05. (page 1)

⇨ Each year, approximately 50 million tonnes of waste comes from industrial sources, such as the food, drink and tobacco industries' production processes. (page 1)

⇨ Landfilling and incinerating waste can place pressures on the environment, e.g. the leaching of nutrients, heavy metals and other toxic compounds from landfills, or the emission of greenhouse gases from landfills and toxins by incinerators. (page 2)

⇨ The UK produces more than 434 million tonnes of waste every year. This rate of rubbish generation would fill the Albert Hall in London in less than two hours. (page 4)

⇨ Every year UK households throw away the equivalent of 3½ million double-decker buses (almost 30 million tonnes), a queue of which would stretch from London to Sydney (Australia) and back. (page 4)

⇨ 64% of the UK's municipal waste was sent to landfill in 2005/06. 27% was recycled or composted, and 8% was incinerated. (page 6)

⇨ We are producing one million tonnes of electrical wreckage annually, a volume that is rising by 5 per cent year on year. (page 8)

⇨ Britain has dedicated an area the size of Warwick – 109 square miles – exclusively to landfill and will run out of space for dumping in less than a decade if current trends continue. (page 10)

⇨ In the UK we are throwing away one-third of the food we buy. That's like one in three bagfuls of food shopping going straight in the bin. (page 11)

⇨ Up to 38 per cent of packaging in a regular household shopping basket cannot be recycled. (page 12)

⇨ Around 13bn plastic bags are given free to UK shoppers every year. The bags can take between 400-1,000 years to break down, and like all forms of plastic they do not biodegrade. Instead they photodegrade, breaking down into smaller and smaller toxic bits that contaminate soil, waterways and oceans, entering the food chain when ingested by animals. (page 13)

⇨ Nearly half the population (48%) admit to dropping litter. (page 15)

⇨ Over 69,000 animals were killed or injured by litter last year in Britain. (page 16)

⇨ Over 30% of an average household bin can be composted at home, from vegetable peelings and teabags, to egg boxes and shredded paper. (page 19)

⇨ The unreleased energy contained in the average dustbin each year could power a television for 5,000 hours. (page 22)

⇨ Over 1.8 million tonnes of old vehicles are thrown away in the UK each year. (page 23)

⇨ There are over 50 different types of plastics. (page 24)

⇨ The energy it takes to make one new aluminium can is enough to make 20 recycled ones. (page 27)

⇨ Every year over 1 million computers end up in our landfill sites. At the moment less than 20% of old computers are recycled. (page 29)

⇨ Recycling increased by 27% across all regions between 2002 and 2006, with households in the east of England recording the highest figure of 34%, almost double the rate of four years previously. London and the north-east recycled the lowest proportion. (page 31)

⇨ The government has set a target for 40% of waste to be recycled by 2010. (page 32)

⇨ Restoring landfill sites by turning them into green space, such as woodland, parkland or farmland, is now possible, new research has shown. (page 33)

⇨ If household waste output continues to rise by three per cent a year, the cost to the economy will be £3.2 billion and the amount of harmful methane emissions will double by 2020. (page 34)

⇨ Currently, landfill sites take up a little over one-tenth of one per cent of the United Kingdom, Richard Wellings of the Institute of Economoc Affairs has said. (page 35)

⇨ Although 57% of the population agrees it is right for retailers to start charging for plastic shopping bags, 52% of people feel they should be able to make up their own minds about this issue and not have it forced on them by retailers wanting to charge. 35% feel it is wrong for the government to ban retailers from giving them away free. (page 36)

GLOSSARY

Biodegradable
Materials that can be broken down by air, water and bacteria, and so have less negative environmental impact than non-biodgradeable materials. Most organic material, such as food and paper, is biodegradable, whereas products such as plastic bottles may take thousands of years to break down – which means it is important to reuse or recycle these non-biodegradable items to avoid them going into landfill.

Composting
Compost is formed from decomposing organic waste which is broken down by the bacteria, insects and animals in the soil. Vegetable peel, fruit, paper and leaves are amongst the materials that can be composted rather than thrown away. Once the material is composted it can be used as fertiliser in the garden.

EU Landfill Directive
The aim of this directive is to decrease the amount of waste sent to landfill and the negative effects of landfill on the environment. The directive requires EU member states (which include the UK) to decrease the amount of biodegradable material they send to landfill to 35% of 1995 levels by 2016.

Energy from waste
Energy from waste or 'waste to energy' refers to the process of burning waste in order to generate electricity. The Institute of Civil Engineers estimates that the UK could generate 17% of the electricity it needs from waste.

Fly-tipping
Term used to describe illegally dumping waste in an area of land that is not an authorised landfill site. People may fly-tip to avoid paying disposal fees or simply through laziness, but if they are caught they can be fined up to £20,000 and/or be imprisoned for six months.

Hazardous waste
Any waste that could be a serious danger to humans or other living organisms, such as chemicals, clinical waste or fuel. Hazardous waste must be disposed of separately to other waste.

Household Waste Recycling Act 2003
This act set out a target for all English waste collection authorities to offer households in their area a doorstep collection of at least two types of recyclable waste by 2010. 90% of households in England currently have this.

Incineration
A method of disposing of waste by burning it. About 9% of UK waste is currently incinerated. Incineration reduces the amount of waste sent to landfill and can convert waste to energy, but there are concerns that it encourages the production of even more waste to keep the incinerators going and that the smoke and ash cause pollution.

Landfill
At a landfill site, waste is deposited into holes in the ground, which are then covered over. In Britain about 70% of waste goes to landfill sites, more than any other country in the EU. It is estimated that we will run out of landfill space in less than a decade if the current disposal rate continues.

Landfill tax
A tax levied on waste sent for disposal at a landfill site, in order to discourage waste production and encourage environmentally-friendly methods of disposal.

Litter
Rubbish that has been dropped in a public place rather than disposed of properly in a bin. Littering is a crime and people who drop litter can be fined £80 on the spot if they are caught.

Municipal waste
Municipal waste refers to all household, commercial and industrial waste.

Recycling
The process of turning something considered waste into a new product. Recycling used materials reduces consumption of natural resources, saves energy and reduces the amount of waste sent to landfill.

The three Rs
The three Rs of rubbish are to Reduce, Reuse and Recycle, (sometimes called the waste hierarchy).

Waste
Anything that is no longer of use and is thrown away. The UK produces more than 434 million tonnes of waste every year.

Waste Electrical and Electronic Equipment (WEEE) directive
This directive makes the manufacturers of electrical goods responsible for financing the collections, treatment, recovery and environmentally-sound disposal of those goods.

Zero waste
A situation in which all waste is either reduced, re-used or recycled, so there is nothing left of which to dispose.

INDEX

Additional Resources

Other Issues *titles*

If you are interested in researching further some of the issues raised in *Waste Issues*, you may like to read the following titles in the **Issues** series:

⇨ Vol. 156 *Travel and Tourism* (ISBN 978 1 86168 443 1)

⇨ Vol. 151 *Climate Change* (ISBN 978 1 86168 424 0)

⇨ Vol. 146 *Sustainability and Environment* (ISBN 978 1 86168 419 6)

⇨ Vol. 134 *Customers and Consumerism* (ISBN 978 1 86168 386 1)

⇨ Vol. 119 *Transport Trends* (ISBN 978 1 86168 352 6)

⇨ Vol. 97 *Energy Matters* (ISBN 978 1 86168 305 2)

⇨ Vol. 76 *The Water Crisis* (ISBN 978 1 86168 265 9)

For more information about these titles, visit our website at www.independence.co.uk/publicationslist

Useful organisations

You may find the websites of the following organisations useful for further research:

⇨ **BusinessGreen:** www.businessgreen.com

⇨ **Campaign to Protect Rural England:** www.cpre.org.uk

⇨ **Defra:** www.defra.gov.uk

⇨ **Economic and Social Research Council:** www.esrc.ac.uk

⇨ **edie:** www.edie.net

⇨ **Encams:** www.encams.org

⇨ **Energy Saving Trust:** www.energysavingtrust.org.uk

⇨ **New Statesman:** www.newstatesman.com

⇨ **Recycle More:** www.recycle-more.co.uk

⇨ **Recycle Now:** www.recyclenow.com

⇨ **Recycling Guide:** www.recycling-guide.org.uk

⇨ **Resource Futures:** www.resourcefutures.co.uk

⇨ **rubbish.co.uk:** www.rubbish.co.uk

⇨ **Waste Online:** www.wasteonline.org.uk

⇨ **WRAP:** www.wrap.org.uk

The publisher is grateful for permission to reproduce the following material.

While every care has been taken to trace and acknowledge copyright, the publisher tenders its apology for any accidental infringement or where copyright has proved untraceable. The publisher would be pleased to come to a suitable arrangement in any such case with the rightful owner.

Chapter One: Our Throwaway Society

Waste in the UK, © Economic and Social Research Council, *Rubbish*, © Guardian Newspapers Ltd, *Wacky waste facts*, © Waste Online, *Waste management*, © rubbish.co.uk, *Consumer adultery*, © New Statesman, *Wasteful Britain: the 'dustbin of Europe'*, © Telegraph Group Ltd, *Wasted food now costs UK homes £10 billion*, © WRAP, *Scale of packaging waste problem*, © Crown copyright is reproduced with the permission of Her Majesty's Stationery Office, *Q&A: plastic bags*, © Guardian Newspapers Ltd, *Carrier bags*, © Crown copyright is reproduced with the permission of Her Majesty's Stationery Office, *Litter*, © ENCAMS, *The problem with litter*, © Campaign to Protect Rural England, *British waste adds to crisis across China*, © Guardian Newspapers Ltd.

Chapter Two: Waste Solutions

Waste at home, © Waste Online, *Tips to reduce waste*, © Energy Saving Trust/WRAP, *A strategy to cut waste*, © Crown copyright is reproduced with the permission of Her Majesty's Stationery Office, *New powers needed to tackle litter louts*, © Crown copyright is reproduced with the permission of Her Majesty's Stationery Office, *Government wants us to recycle on the go*, © edie, *Recycling facts and figures*, © Recycling Guide, *Recycling tips*, © Recycle More, *The machine that sorts out household rubbish*, © 2008 Associated Newspapers Ltd, *Recycle and reuse*, © Resource Futures, *Steps to successful home composting*, © Recycle Now, *How green are we?*, © Guardian Newspapers Ltd, *Hazardous landfill waste falls*, © Incisive Media 2008, *Landfill sites have a green future*, © Crown copyright is reproduced with the permission of Her Majesty's Stationery Office, *Recycling is not enough – we need to consume less*, © Economic and Social Research Council, *Why recycling isn't really saving the planet*, © Institute of Economic Affairs/Yorkshire Post, *Government ready to act on plastic bags*, © Crown copyright is reproduced with the permission of Her Majesty's Stationery Office, *Consumers oppose nanny state on plastic bags*, © Ipsos MORI, *Waste exports*, © Crown copyright is reproduced with the permission of Her Majesty's Stationery Office, *Time to waste – tackling the landfill challenge*, © New Local Government Network, *Renewable energy from rubbish is possible*, © Telegraph Group Ltd.

Photographs

Flickr: page 36 (Brandi Tressler).
Stock Xchng: pages 3 (Ted Rosen); 22 (Sanja Gjenero); 24 (resignent); 27 (Alessandro Paiva); 29 (Sophie).

Illustrations

Pages 1, 16, 34: Angelo Madrid; pages 4, 17, 39: Don Hatcher; pages 7, 30: Bev Aisbett; pages 12, 19, 33: Simon Kneebone.

Editorial and layout by Claire Owen, on behalf of Independence Educational Publishers.

And with thanks to the team: Mary Chapman, Sandra Dennis, Claire Owen and Jan Sunderland.

Lisa Firth
Cambridge
September, 2008